BEYOND THE BORDER

Beyond the Border

Tensions across the Forty-Ninth Parallel in the Great Plains and Prairies

Edited by

KYLE CONWAY AND TIMOTHY PASCH

McGill-Queen's University Press
Montreal & Kingston · London · Ithaca

© McGill-Queen's University Press 2013

ISBN 978-0-7735-4130-6 (cloth)
ISBN 978-0-7735-4131-3 (paper)
ISBN 978-0-7735-8862-2 (ePDF)
ISBN 978-0-7735-8863-9 (ePUB)

Legal deposit second quarter 2013
Bibliothèque nationale du Québec

Printed in Canada on acid-free paper that is 100% ancient forest free (100% post-consumer recycled), processed chlorine free

This book has been published with the help of funding from the Office of the Vice President for Research and Economic Development at the University of North Dakota and from the conference grant program at Foreign Affairs and International Trade Canada.

McGill-Queen's University Press acknowledges the support of the Canada Council for the Arts for our publishing program. We also acknowledge the financial support of the Government of Canada through the Canada Book Fund for our publishing activities.

Library and Archives Canada Cataloguing in Publication

Beyond the border: tensions across the forty-ninth parallel in the Great Plains and Prairies / edited by Kyle Conway and Timothy Pasch.

Based on papers presented at a conference held at the University of North Dakota Institute for Borderland Studies, June 2010.
Includes bibliographical references and index.
ISBN 978-0-7735-4130-6 (bound). – ISBN 978-0-7735-4131-3 (pbk.)
ISBN 978-0-7735-8862-2 (ePDF). – ISBN 978-0-7735-8863-9 (ePUB)

1. Borderlands – Social aspects – Prairie Provinces – Congresses.
2. Borderlands – Social aspects – Great Plains – Congresses. 3. Borderlands – Political aspects – Prairie Provinces – Congresses. 4. Borderlands – Political aspects – Great Plains – Congresses. 5. Prairie Provinces – Social conditions – Congresses. 6. Great Plains – Social conditions – Congresses. 7. Group identity – Prairie Provinces – Congresses. 8. Group identity – Great Plains – Congresses. I. Conway, Kyle, 1977– II. Pasch, Timothy, 1974–

FC186.B49 2013   303.48'2712078   C2013-901067-X

This book was typeset by Interscript in 10.5/13 Sabon.

Contents

Figures   vii

Acknowledgments   ix

Introduction: Paradoxes of the Border   3
Kyle Conway and Timothy Pasch

THE MEDIATED BORDER

1 The Borders of Cultural Difference: Canadian Television and Cultural Identity   29
Serra Tinic

2 The Canadian Sitcom and the Fantasy of National Difference: *Little Mosque on the Prairie* and English-Canadian Identity   39
Christopher Cwynar

3 The Flow of Amusement: The First Year of Moving Pictures in the Red River Valley   71
Paul Moore

THE POLITICAL BORDER

4 "Shutting Down the Snake Ranch": Battling Booze at the BC Border, 1910–14   93
Brandon Dimmel

5 International and Domestic Pressures on the Governance of the St Mary and Milk Rivers   113
Michelle Morris

6 Water and Political Relations between the Upper Plains States and the Prairie Provinces: What Works, What Doesn't, and What's All Wet   133
Paul R. Sando

THE NATIVE BORDER

7 Border Studies and Indigenous Peoples: Reconsidering Our Approach   153
Zalfa Feghali

8 Navigating the "Erotic Conversion": Transgression and Sovereignty in Native Literatures of the Northern Plains   170
Joshua D. Miner

9 The Anishnaabeg of Bawating: Indigenous People Look at the Canada-US Border   199
Phil Bellfy

Conclusion: Beyond the Paradoxes of the Border   223
Kyle Conway and Timothy Pasch

Contributors   239

Index   243

Figures

3.1 The Red River Valley   72
3.2 Red River Valley theatrical circuit   77
3.3 Cartoon depicting the men of Grand Forks, ND, "studying" the atrocities at the Museum of Anatomy   83
4.1 Peace Arch unveiling, White Rock, BC, 6 September 1921   94
4.2 The St Leonard Hotel, White Rock, BC   99
4.3 Peace Arch, White Rock, BC   108
5.1 Map of the St Mary and Milk rivers   114
5.2 Actor influence on Alberta   117
6.1 The Souris River Basin   138
6.2 Drainage context of Devils Lake   141
6.3 The glacial geology of the Devils Lake basin   141
6.4 Proposed Red River diversion   144
9.1 Satellite image of the Canada-US border between Garden River First Nation and Sugar Island   207
9.2 Treaty of Ghent boundary, upper termination line, 1828   210
9.3 Treaty of Ghent boundary, lower termination line, 1828   211
9.4 Detail of Schoolcraft's 1837 map of Michigan   212
9.5 Transcription of a portion of Schoolcraft's 1837 map of Michigan   213
9.6 Portion of the 1847 Mitchell map of Michigan   214
10.1 An unmanned aerial vehicle deployed for border patrol   234

Acknowledgments

It is a well-established convention to observe that all books are inherently collaborative. In a collection such as this, that collaboration is even more pronounced, and to begin, we would like to thank everyone who attended or presented a paper at "The Great Plains, the Prairies, and the US/Canadian Border," the conference out of which this book grew. We were impressed and pleased with the quality of the exchanges, especially from those scholars who travelled great distances to Grand Forks, North Dakota, our little town in the middle of the continent. There would have been no conference (and no book) without their generous contributions. We are grateful for their patience and attention to detail, as well as for their insight.

We would also like to thank the many organizations that provided funds to make it possible for the University of North Dakota Institute for Borderland Studies to host the conference, including Foreign Affairs and International Trade Canada (through its conference grant program) and the Consulate General of Canada in Minneapolis, especially Amy McBeth, Courtney Selstad, Jennifer Kay, Christine Davis, and Consul General Martin Loken. We could not have asked for more professional or friendly colleagues. Thanks are due also to the University of Manitoba, which provided funds through its Canadian studies program. Not least of all, we want to thank the University of North Dakota Office of the Vice President for Research and Economic Development, in particular Phyllis Johnson and Barry Milavetz, and the University of North Dakota School of Graduate Studies, in particular Joey Benoit. These offices also provided funds that helped pay for the publication of this book. Finally, thanks to the Greater Grand Forks Convention and Visitors Bureau, in particular Deb Stewart, for all the on-the-ground help. We hope that we have made good on the considerable investment of time and resources made by all of these organizations.

We would also like to acknowledge how indebted we are to the people at the University of North Dakota who laid the groundwork for the Institute for Borderland Studies before either of us even arrived here. They include Joan Hawthorne and Greg Weisenstein, both members of the administration who worked to establish the institute. Above all, however, we would like to thank James Mochoruk, Doug Munski, and Virgil Benoit, who are also the people most responsible for the vibrant and growing Canadian studies program at the university. Without their years of hard work, and, more to the point, without their friendship and guidance, the conference and this book would not have been possible.

Thanks also to the editorial staff at McGill-Queen's. Kyla Madden's prompt feedback and expert guidance at each stage of the book's production were invaluable, and the book is stronger thanks to her attention to detail.

KYLE'S ACKNOWLEDGMENTS

I would like to thank my co-editor Tim for graciously stepping in at the last minute (when the birth of my daughter called me away) and running the conference itself. He has been the model of professionalism and collegiality, not to mention a kind and steadfast friend. He has also demonstrated time and again the value of collaboration, and it is for that reason that this book is more than a sum of its parts.

Most importantly, I would also like to thank my beautiful wife Kristi and my daughter Eleanor, whose birth the day before the conference made it even more joyous than I had anticipated.

TIM'S ACKNOWLEDGMENTS

I could never have imagined when Kyle stepped into my office one fine afternoon that this journey could have been so infinitely rewarding. Kyle's keen attention to detail, scholarly rigour, and professional and personal goodwill has and continues to positively inspire and impact my scholarship and worldview. With the utmost respect, I would also like to thank a variety of individuals who have additionally catalyzed my research and significantly assisted me in ways beyond measure: Dr Anthony Chan, my doctoral chair, friend, and professor emeritus of the UW and UOIT, Nadine Fabbi, associate director of Canadian studies at the University of Washington; the incomparable Chris Kirkey and André Senecal of Project CONNECT at SUNY Plattsburg, where I met Kyle for the first time; Donat Savoie for his mentorship; the Avataq Cultural Institute in Inukjuaq; and our Canadian studies colleagues at UND. Thank you; nakurmiik, et merci mille fois mes amis. Finally, it would be remiss to omit my greatest supporters: my lovely wife Saori and our sons Kai and Riku. Domo Arigatou!

BEYOND THE BORDER

Introduction: Paradoxes of the Border

KYLE CONWAY AND TIMOTHY PASCH

The Canada-US border serves a paradoxical function: it separates two countries, even as it sutures them together. As a result, weird things happen there. The border interrupts space and marks a break between one discrete place and another. We experience this each time we cross it, stop at the inspection station, and hand over our passports. The space we pass through is not continuous.

The border also interrupts time, although the break is perhaps more abstract. To cross the border is to leave one temporal frame of reference and national timeline, replete with its own history and sense of order, and arrive in another. It is to enter into a new relationship with the past, and consequently with the present and the future. At the border, time is no more continuous than space.

These breaks happen at borders marked by trauma, such as the Mexico-US border, which Gloria Anzaldúa (2007, 25) describes as "*una herida abierta*," an open wound, "where the Third World grates against the first and bleeds." They also happen at borders that "work," such as the Canada-US border, long described as the world's longest undefended border (rebranded as the world's longest *secure* border after the attacks of 11 September 2001). They even happen in places sometimes overlooked as staid, such as the northern Great Plains and the Prairies, which may appear to people on the coasts as fly-over country.

That staid and potentially overlooked border was the focus of the conference that led to this book. In June 2010, the University of North Dakota Institute for Borderland Studies hosted "The Great Plains, the Prairies, and the US-Canadian Border," a two-day conference that aimed to bring scholarly attention to a frequently neglected portion of the North American continent. It was one of a growing number of conferences about the Canada-US border, which have produced a fair number of books in the last decade. Much of the focus has been on the western border region, in books such as Sterling Evans's *Borderlands of the American and Canadian Wests* (2006) and Elizabeth Jameson and Sheila McManus's *One Step over the Line* (2008). The Great Lakes region has also received considerable attention, in books such as John Bukowczyk, Nora Faires, David Smith, and Randy Widdis's *Permeable Border* (2005) and Karl Hele's *Lines Drawn upon the Water* (2008). In these conferences and books, historians and geographers lead the way and provide a rich array of approaches and a careful consideration of the nature of the different, often-contradictory factors that set one region apart from another.

When we issued our call for papers, our plan was to follow the lines this recent work suggested. At the same time, we also planned to address a sentiment we frequently encountered, living in the centre of the North American continent, that life is somehow less interesting here than in cities such as Toronto or New York. What we found, however, was that the two aspects of that plan were far more complicated than they first appeared. For one thing, what exactly constitutes the Great Plains region (and its corresponding region on the other side of the forty-ninth parallel) is difficult to pin down. For another, many previous approaches were inadequate to the task of providing a more complete understanding of fly-over country.

In the first instance, the nature of the region we wanted to discuss was complicated by the idea of region itself, which has fallen in and out of favour since it was first used in the 1930s. At that time, social scientists, especially physical geographers, thought that "regions emerge, first of all, from certain physiographic uniformities" from which "economic, social, and cultural relationships" result, according to Frederick Luebke (1984, 20). Such environmental determinism came under attack in the 1960s, by which time "physical geographers became less certain of their large categories, less magisterial in their regional delineations ... The Great Plains, described so confidently by [Nevin] Fenneman in 1931 [in *Physiography of Western United States*], literally disappeared from some textbooks of a later generation" (23–4). As a consequence of such disciplinary back-and-forth, the boundaries of the Great Plains remain open for debate – they are bordered on the west by the Rocky Mountains, but their eastern limits are far less clear. Further complicating matters, as described below, is the fact that decisions that affect the Great Plains and Prairies are often made in other places, such as when policy-makers in one region make decisions about rivers that flow into another.

In the second instance, historians' approaches were inadequate for providing a deeper understanding of the role of the border in the centre of the continent. Thomas Isern and R. Bruce Shepard (2006, xxxi–xxxii) describe three approaches that have characterized historians' study of the Canada-US borderlands. First is the continentalist approach, which "begins with the assertion of a regional integrity that crosses national lines." Second is the comparative approach, which draws on the "classic comparison-contrast format" where scholars "sort out the similarities and differences and generally emphasize the differences." Although these approaches provide a useful framework for examining a region's history, they are less helpful for considering how the paradoxes of the border play out for the people who live there. In that respect, the third approach, which Isern and Shepard call the "borderlands" approach, supplies more useful leads. It builds on the theories of post-colonialism developed in the fields of literature and cultural studies and "dwells geographically on lands and peoples physically adjacent to the international boundary or in some way associated with it." It emphasizes the idea, in Karl Hele's (2008, xv) words, that "borders are lived experiences."

The borderlands approach is in line with certain contemporary approaches in geography. Although geographers have investigated borders since the late nineteenth century (when borders were growing in political importance as European powers established their empires), their investigation of the Canada-US border has been a relatively recent phenomenon (Konrad and Nicol 2008, 34–7). Instead, their focus has frequently been theoretical and related to questions along these lines: will borders (and the limits of political sovereignty that they signal) wither away in an era of globalization and boundary permeability (Blake 2005)? Similarly, how do states (and other political entities) manage the "interplay and interdependence between individuals' incentives to act and the surrounding structures (constructed social processes that contain and constrain individual action ...) that determine the effectiveness of formal border policy" (Brunet-Jailly and Dupeyron 2007, 2)? In this vein, David Newman and Anssi Paasi (1998, 191) identify four major themes that characterize contemporary boundary studies: "1) the suggested 'disappearance' of boundaries; 2) the role of boundaries in the construction of sociospatial identities; 3) boundary narratives and discourse; and 4) the different spatial scales of boundary construction." The essays in this book address all of these themes.

The papers that speakers presented at the conference (which took place in Grand Forks, North Dakota, about ninety miles south of the border) drew on a wide range of disciplines and reflected the idea of lived experience. As the conference progressed, however, a deeper undercurrent began to take form. In particular, presenters spoke of ways that our interactions with and at the border bring to light contradictions and challenges posed by globalization. These challenges resembled those described by political philosopher Jürgen Habermas (2001, 62, 68–9), who observes that a wide range of phenomena, including ecological degradation, organized crime, and increased capital mobility, challenge the modern state's ability to "[maintain] sovereignty over a determinate geographical territory." In the northern Great Plains and the Prairies, these phenomena take a different form – for instance, media that cross the border, rivers that cross the border, and native communities that cross the border – but their implications for states exercising administrative powers over a "determinate geographical territory" are similar.

We used Habermas's observation as one starting point when organizing the essays in this book, which collectively trace an evolution in notions of the border. In the very first essay, Serra Tinic presents one perspective, that of the US State Department, which sees the border as a material, institutional line whose ontological status goes unquestioned: the border is merely given. But that status is quickly questioned as the border comes to appear as a porous line before being reduced in the final chapters to an illusion of history. To be sure, it is not a question of whether a line separates the United States and Canada (see Newman and Paasi 1998, 191–3), but of the effort necessary to ensure that line's continued existence. An index of the border's fragility, we argue, is the force required to maintain it: strong borders are strong because they go unchallenged, while fragile borders are fragile because they are drawn into question. Thus, the more fragile a border, the more effort that is required to enforce it. By the end of the book, we must reconsider those arguments that had previously seemed inviolate through a new lens, which may reveal creative and innovative possibilities for

Introduction

enhancing the desirable porousness of the border, while still warding against those threats that both sides can vehemently agree on.

We have grouped the essays by theme and divided the book into three sections: "The Mediated Border," "The Political Border," and "The Native Border." The object of inquiry defines these sections, but it would have been equally possible to trace other themes that intersect the different chapters.[1] In the first section, *mediated* has two distinct senses: "As portrayed by entertainment media" and "produced and known through discourse." Although the first sense is the most immediately apparent in the essays in question – the first two (by Serra Tinic and Christopher Cwynar) discuss depictions of the border in programs such as *Little Mosque on the Prairie* and the third (by Paul Moore) discusses early film distribution networks in the region – it is the second sense that is the more important. These essays emphasize ways in which the border is produced and reproduced through various discourses, and consequently, what appears first as given – the border as an institutional, even physical line – is undermined by the fact that it is ultimately dependent on the mental boundaries that governments, citizens, and the media construct.

Political in the second section, like mediated in the first, also has two distinct facets: "Separating more than one political entity or *polis*" and "subject to contestation and negotiation." The essays in this section examine the border in both senses of the political in relation to issues of water management, which fall under the purview of state, provincial, and national governments that must negotiate with each other to resolve their conflicting views and goals. The Great Plains and Prairies are home to a number of rivers, including the St Mary, Souris,[2] and Red rivers, which criss-cross the border and empty eventually into Lake Winnipeg in Manitoba. Michelle Morris writes about the influence wielded by governmental, non-governmental, and commercial entities with competing interests in the management of the St Mary River and the Milk River (a tributary of the Missouri). Paul Sando examines the different approaches taken to manage cross-border flows of the Souris and Red rivers, as well as the potential for flooding from Devils Lake, whose controversial and as-yet-unrealized outlet would drain eventually into the Red River. In these essays, the border is a porous line, and its porousness presents another challenge to its ontological status – it fails in doing precisely what it is meant to do, namely mark a division between two political entities, a failure that different levels of government feel with increasing acuity as they cede their ability to shape how regulation of the border takes place.

Native, like mediated and political, also has two meanings: "As seen by indigenous peoples" and "originary." These two meanings raise a bigger question, which the essays in the final section seek to answer: in what sense can borders be native to a place – that is, in what sense can they originate in a place – and in what sense must they be imposed? Zalfa Feghali frames this question by asking how scholars might study the border in the Great Plains and Prairies using tools developed by scholars looking at the Mexico-US border. Joshua Miner asks what native literature written by Great Plains authors reveals about the discursive reproduction of the border. Although he addresses largely the same themes as Tinic and Cwynar, he goes further and argues that stories told from native points of view "destabilize the national narratives that have subordinated and marginalized them for so long" (171). This destabilization draws the border's ontological status further into question – the line that defines the border loses its legitimacy when we see it merely as the product of colonial imposition.

We have included two additional essays in this collection that, although they are concerned with regions outside of the Great Plains and Prairies, help to complete the thematic arc of the book while also lending valuable complementary insights. The first, included in the section on the political border, is Brandon Dimmel's article on the conflict between residents of the border towns of Blaine, Washington, and White Rock, British Columbia, over the northward flow of sewage and the southward flow of liquor in the early 1910s. It provides a valuable point of comparison to Moore's chapter on film distribution (in its observations about the role of geography and transportation infrastructure in the discursive construction of the border) and Morris's and Sando's chapters on river management (in their discussions of the residents' appeals to distant national capitals to settle local disputes). The second, which concludes the section on the native border and the book itself, is Phil Bellfy's description of the Canada-US border as seen through the eyes of the indigenous peoples living in the Great Lakes region. In his essay, he deconstructs many a priori concepts and argues that because of the illegitimacy of its formation, the history of treaties between the United States and Britain (later Canada), and the vicissitudes of nature, the border

– from the point of view of those against whom it was imposed – is nothing more than an illusion. The questions Bellfy poses, we conclude, not only call upon us to reconsider the chapters that precede his, but also constitute a calling into question of certain commonly accepted tenets of borderland studies.

THE MEDIATED BORDER

Historically, Canada's concern about media crossing the border has been a concern about cultural imperialism. At its simplest, it has reflected a fear that English Canadian culture cannot survive the onslaught of US media, especially broadcasting (see Rutherford 1993). After all, the electromagnetic spectrum that carries radio waves respects no boundaries. A more nuanced account, however, takes the form of a synthesis of this cultural imperialism thesis and the Frankfurt School's culture industries argument, with its focus on how the logics of industry, including the constant need to attract and retain audiences, make it impossible to transcend the formulaic and predictable. Max Horkheimer and Theodor Adorno (1972), in an essay originally published in 1947, identify an insidious effect of popular media: using a combination of flattery and deception, the companies that made movies and jazz records lulled consumers into such complacency that they mistook formula for originality and, as a consequence, could no longer recognize genuine originality that might have led to socially redemptive ends. Dallas Smythe (1981) extends their argument while adding a new dimension: the media industries in post-Second World War Canada were controlled largely by US companies, and by controlling content, US media companies also controlled how Canadians interpreted the world.

The essays in this book present a new set of concerns, related to but qualitatively different from those of the past. The idea that US media might somehow "infect" English Canada and strip it of its identity has lost its urgency as scholars have investigated other factors shaping English Canadian identity.[3] Not all of these factors relate to the media. Charles Taylor (1993, 158–61), for instance, names a number of values Canadians draw on to distinguish themselves from Americans, including the maintenance of law and order, the collective distribution of resources, and fiscal equalization between regions, as well as political institutions that serve as touchstones for English Canadian identity, including multiculturalism and the Charter of Rights and Freedoms. In his analysis, English Canadian identity is constructed, not imposed, by the United States, although it is frequently constructed as a form of negative identity with the United States acting as a foil. In this respect, the essays in this section take an approach similar to the one "in recent social and cultural theory [where] the idea of boundary refers increasingly to the social and symbolic construction of boundaries between social collectivities" (Newman and Paasi 1998, 194). But they take the metaphor of border at face value, turning our attention back to state boundaries by considering the relationship between identity and the discursive production of the Canada-US border.

The arc of this book, defined by the progressively deeper investigation of the border's ontological status, begins here. Perhaps no organization takes the border's existence as a given in quite the same way as the US departments of State and Homeland Security, charged respectively with diplomacy and border security enforcement. Both departments are premised on the border's status as an institutional given. Both have also taken notice of recent trends in Canadian television programming, especially by the Canadian Broadcasting Corporation (CBC). In a diplomatic cable from the American Embassy in Ottawa, a US State Department official wrote with some alarm:

> When American TV and movie producers want action, the formula involves Middle Eastern terrorists, a ticking nuclear device, and a (somewhat ironically, Canadian) guy named Sutherland. Canadian producers don't need to look so far – they can find all the action they need right on the US-Canadian border. This piece of real estate, which most Americans associate with snow blowing back and forth across an imaginary line, has for the past three weeks been for Canadian viewers the site of downed rendition flights, F-16 bombing runs, and terrorist suspects being whisked away to Middle Eastern torture facilities. (United States 2008)

The offending program, as Tinic writes in the essay that opens the section on the mediated border, was an hour-long drama called – fittingly – *The Border*. The official's complaint concerned the unflattering portrayal of US agencies. In *The Border*, the United States is represented by "an arrogant, albeit stunningly attractive female [Department of Homeland Security] officer," played by Sofia Milos, "sort of a cross between Salma Hayek and Cruella De Vil" who walks around uttering "such classic lines as, 'Who do you think provides the muscle to protect your fine ideals?' and 'You would have killed him. Let the American justice system do it for you'" (United States 2008). The comedy *Little Mosque on the Prairie*, as Tinic explains, addresses a different section of the border, this time in Saskatchewan, but the official's concern remains the same: "A December 2007 episode portrayed a Muslim economics professor trying to remove his name from the No-Fly-List at a US consulate. The show depicts a rude and eccentric US consular officer stereotypically attempting to find any excuse to avoid being helpful. Another episode depicted how an innocent trip across the border became a jumble of frayed nerves as Grandpa was scurried into secondary by US border officials because his name matched something on the watch list" (United States 2008).

The official's complaint is revealing on several levels. As Tinic points out, "The American cultural 'other' permeates the broader Canadian cultural imagination and ... domestic broadcasting policy has long stressed that a televisual border of difference needed to be fostered if Canadians were to see themselves as a unique imagined community. Ironically, the state department's alarm is evidence that, in recent years, the CBC has fulfilled its mandate" (37). More revealing, however, is the subtle slippage that occurs between "this piece of real estate" described by the state department official – a geographically locatable institutional line – and the "televisual border of difference" Tinic alludes to – an imagined line that exists only as a shared idea.

That slippage is the first point where the border's ontological status is called into question. The fact that the imagined border *must* be reinforced discursively, as the US State Department official's complaint strongly suggests, draws into question the givenness of the border as a line of demarcation. Christopher Cwynar demonstrates as much in his chapter, "The Canadian Sitcom and the Fantasy of National Difference." He extends Tinic's analysis of *Little Mosque* by considering the processes by which the discursive production of the border takes place. *Little Mosque* is characterized by an ambivalence between those aspects of a program that mark it as Canadian and those that evoke ideas of someplace else, especially the United States. This ambivalence is built into the program on at least two levels. First, the show is a clear product of the Canadian multicultural, "mosaic" discourse – one of the touchpoints of English

Canadian national identity – that stands in opposition to the "melting pot" discourse Canadians see as characteristic of debate in the United States. The show is also a product of the CBC – earnest, informative, and careful to address more issues than just that of Islam in North America. It is a regional program, for instance, and fulfills the CBC's mandate to "reflect Canada and its regions to national and regional audiences" (Broadcasting Act, 1991, sec. 3[1][m][ii]). It presents an image of rural Saskatchewan far outside the Toronto-Ottawa-Montreal corridor usually featured in Canadian media.

There is a tension between its "Canadianness" and the sitcom genre itself, which Cwynar argues is typically associated with US television. Borrowing from Mary Jane Miller's (1993) analysis of ways in which Canadian producers have appropriated and reworked US genres, he argues that *Little Mosque* "inflects" the sitcom "in a manner that alludes to both established themes in Canadian nationalism and the CBC's core values" and by incorporating "informational or educational content" (49–50). The second point of ambivalence relates to *Little Mosque*'s use of satire (which serves as a vehicle for cultural critique) and parody (which serves to drive comedic action). Parody relies on viewers' knowledge of other programs, especially those produced in the United States, which leads Cwynar to conclude that satire and parody "reaffirm the importance to Canadians of a shared repository of mass media texts that has largely been produced by the American cultural industries. In other words, if the satirical elements provide opportunities for critical distinction along national lines, the straightforward parodies reassert the importance of a shared Anglo-American popular cultural heritage that the United States has overdetermined" (53). Thus the "televisual border of difference," the imagined border that serves as a bulwark in support of Canadian national identity, is entirely dependent on the very thing it is meant to deny: the influence of US culture on Canadian identity.

This paradox extends beyond program content and the textual realm, however, as Paul Moore shows in his historical analysis. Moore looks at the "structural and cultural institutions" (73) that shaped early film distribution, including policy (the standards of delivery of electricity and the east-west orientation of Canada's transportation infrastructure), geography (the north-south orientation of the Red River), and industry (the organization of early film distribution). He frames the border in terms borrowed from Louis

Althusser (1971). In Althusser's Marxist philosophy, the repressive state apparatus (or RSA) "contains: the Government, the Administration, the Army, the Police, the Courts, the Prisons, etc ... Repressive suggests that the State Apparatus in question 'functions by violence' – at least ultimately (since repression, e.g. administrative repression, may take nonphysical forms)" (142–3). The ideological state apparatuses (or ISAs) are institutions such as schools, churches, or the media that function by ideology. Because their ideology is that of the ruling class, they support the RSA. In fact, if they function well, they obviate the need for an intervention by the army, the police, the courts, or the prisons. If they function well, it is because we accept our place in them reflexively, without an explicit awareness of our act of acceptance.

Moore's recourse to these terms helps clarify the tension identified above between the two valences of "border." The imagined border functions as an ISA, while the RSA (embodied, for example, in the US departments of State and Homeland Security) works to enforce the border as institutional line: "If nationalism is the means of establishing acceptance of an imagined community, then the nation's boundary line becomes partly an imagined – rather than simply an institutional – fact that is continually reproduced in daily practice through a variety of state, economic, and cultural apparatuses" (73).

These apparatuses have taken their shape from a number of sources, and their effects were observable in the distribution patterns of early film technology. On the one hand, the north-south orientation of the Red River, stretching from near Fargo, North Dakota, up to Winnipeg, Manitoba, appears to encourage travel along north-south routes, and, indeed, the earliest distribution did follow this route. American John Cryderman bought the rights to Thomas Edison's Vitascope in March 1896 and collaborated with Canadian Richard Hardie to bring the first screening to the region, a screening that took place in the small border town of Pembina rather than the more populous US cities to the south because Pembina had the requisite power supply. The population centres in this region were all along the Red River, and they attracted the interest of the "Chicago-based professional shows" (86).

On the other hand, despite what must have been the appeal of relatively close US cities, smaller Canadian distributors (including Hardie, after breaking with Cryderman) deliberately chose east-west routes, thoroughly exploiting their regional markets and using "the nation-building institution of the Canadian Pacific Railway to bring their amusements to the Canadian Northwest Territories, eventually to the British Columbia coast" (87). US distributors also did not venture north of the forty-ninth parallel, in part because the content of their films was too graphic for Canadian tastes. In one case, for example, there was concern that film of a lynching in Texas might be confiscated at the border.

Moore concludes that it was Canadians' "overtly nationalist entertainments ... that turned the border into a barrier" (86). In other words, it was an ideological – rather than repressive – state apparatus that enforced the border's status. It is useful to note, if we follow Althusser's analysis, that when ISAs fail, the RSA steps in to impose order. Earlier, we noted that the border as an institutional line depends on the mental boundaries we construct. Ultimately, if those mental boundaries fail, then the institutional line can be protected only through state force.

THE POLITICAL BORDER

The articles in the section on the mediated border finish by reminding us that in the final instance, borders exist only because the apparatuses of the state enforce them. They are ultimately political, in that they separate one *polis* from another and – because mental boundaries must be continually reproduced – they are subject to negotiation and even contestation. What are the implications, then, of the doubly political nature of the Canada-US border?

Moore's historical analysis raises questions about the influence of institutions of geography (such as the east-west orientation of the Canadian Pacific Railway) on people's mental geography, or the ways they experience space and understand their relationship to it. These questions, which relate to what Newman and Paasi (1998, 191, 197–8) call "the different spatial scales of boundary construction," are a good starting point for considering the political border. They are similar to the questions raised by Brandon Dimmel in the chapter that opens this section. Dimmel takes a historical approach to a conflict pitting the border towns of Blaine, Washington, and White Rock, British Columbia, during the era of prohibition in the United States. Like Moore, he examines the interplay between geography and the imagined relationships that give shape and texture to people's sense of national identity. The border between Blaine and White Rock, he writes, "was not marked by a natural or man-made barrier [but it] did exist" (94). It was, however, rather porous, with people and booze passing from White Rock's St Leonard Hotel into Blaine, which was dry, and waste passing from a fishery in Blaine into White Rock, which wanted to become a resort town. Dimmel provides a sense of how the border interrupts space and how relationships with centres of government geographically removed from local events can be closer than relationships with people who are geographically near but separated by a political border. This distance becomes apparent in the different authorities to which people appealed in their conflict with each other. White Rock, for instance, appealed to the International Joint Commission (IJC), established as a binational commission in 1909 by the Boundary Waters Treaty, to intervene on its behalf. In other words, White Rock residents appealed to an international authority to settle local problems.

The question of national identity has been central to the legitimacy of the democratic nation-state, but the porousness of the border presents a serious challenge. This is Jürgen Habermas's (2001) point in his discussion of the "postnational constellation." He identifies four historical conditions societies have met as they realized themselves as capable of democratic self-control: they have developed "as an administrative state supported through taxation (a); maintaining sovereignty over a determinate geographical territory (b); in the specific form of the nation-state (c), which then democratically developed into a legal and social state (d)" (62). The increasing porousness of borders across the world challenges all of these conditions, with capital mobility, for instance, drawing notions of geographic territory into question and making it more difficult to collect taxes.

Dimmel's discussion of the IJC raises two related questions. What happens when natural resources such as water cross the border? Second, how does changing mental geography – or the way one conceives of and experiences space – influence notions of national identity on which the political entity of the nation-state has historically been premised? The other two chapters in the section on the political border – Michelle Morris's discussion about the Milk and St Mary rivers and Paul Sando's discussion about water issues in North Dakota and Manitoba – present case studies about rivers that consider these questions in more detail. If the permeability of the border suggests its own "disappearance" or decreased relevance (Newman and Paasi 1998, 191) anywhere among these essays, it is here.

Morris's and Sando's essays build on the growing literature about transnational river management, which itself resonates with Habermas's observations about the challenges posed by porous borders. Ines Dombrowsky (2007, 3–4) describes the scope of the phenomenon of transnational rivers, of which, by one count performed in the late 1990s, there are 263: "Approximately 40 percent of the world's population lives in international river basins. They account for an estimated 60 percent of the global freshwater flow and ... cover 45 percent of the earth's land surface." Rob de Loë (2009, 20) lists the range of scholars interested in the phenomenon, including those who work in "transboundary water management, water governance, climate change, water security, and international water law." Such scholars work to address one of the fundamental characteristics of water flow: what happens upstream affects downstream, which forces governments to work together (or not) when upstream and downstream are in different jurisdictions.

Historically, agreements that address border-crossing rivers have been "single-purpose and [have tended] to be confined to specific sections of the respective river basins" (Dombrowski 2007, 13). Increasingly, however, there is a trend toward more integrated approaches that attempt to take a broader look at the different factors (environmental, governmental, economic, and so on) that affect the rivers in question. International benchmarks for effective transboundary water management, set forth in places such as the 1997 United Nations Water Convention, measure the degree to which agreements adopt an integrated approach, protect the larger ecosystem, involve the public, share governance, and permit flexibility and adaptability in their application (de Loë 2009, 20). Such integrated approaches demand, however, "that jurisdictional boundaries be downplayed in the search for collaborative, integrated basin management" (18). In particular, "in an environment of shared or distributed governance, the state ... shares or lets go of some of its authority" (24), much as Habermas points out.

Morris approaches these questions by examining the management of the Milk River, which starts in Montana and is a tributary to the Missouri River, and the St Mary River, which originates in Montana and is a tributary to the Saskatchewan River, which empties eventually into Lake Winnipeg. She examines the different interested parties – the IJC, national governments, state and provincial governments, sectoral players (i.e., those with economic interests such as agricultural players), native groups, and environmental non-governmental organizations (ENGOs) – and uses evidence of their influence in a recent series of negotiations as an index of their importance relative to each other. The IJC, for instance, has been relatively weak because it has no enforcement mechanisms. Other actors with limited influence include native groups such as the Kainah, who were excluded from important bilateral negotiations and whose claims to water rights have gone unrecognized, and ENGOs, whose initiatives have largely been ignored by the governments concerned. Groups with more clout have included the Blackfeet nation, which in contrast to the Kainah has been involved in the policy-making process. The actors with the most clout have represented sectoral interests, especially those related to agriculture, that exert considerable influence, having inspired "Alberta's protection of the 1921 Order on Apportionment, the subsidized rehabilitation of southern Alberta's expansive canal network and irrigation schemes, revisions to the Water Act to allow a water market to develop within the South Saskatchewan River Basin (SSRB), and a feasibility study regarding on and off stream storage on the Milk River" (116).

Paul Sando performs a similar analysis but focuses on efforts to control water flow and flooding in North Dakota and Manitoba. He examines the Souris River (also known as the Mouse River on the US side of the border), Devils Lake (near the town in North Dakota of the same name), and the Red River (which also figures prominently in Moore's chapter). These are all bodies of water whose management necessarily involves multiple state actors. The Souris begins in Saskatchewan before it passes into North Dakota, then back into Manitoba, where it joins the Assiniboine River. Similarly, Devils Lake has risen steadily for the past decade because it has no natural outlet. The rising waters threaten the town of Devils Lake, but a proposed artificial outlet would drain eventually into the Red River and from there into Lake Winnipeg. The Manitoba government has fought such an outlet because of the threat it would pose to Manitoban waters, meaning that Devils Lake residents are subject to the decisions made by a government that is not their own. In this way, water management has a tangible effect on the mental geography of those people affected by the excess of water. When the Red

River rises each spring, for instance, residents of Grand Forks, North Dakota, feel a certain kinship with their Manitoban neighbours as the north-flowing river and its valley become a shared space, even if more formal relationships are governed by officials in Bismarck and Winnipeg, or in Washington and Ottawa.

Throughout Morris's and Sando's chapters, what becomes clear is that the number of stakeholders (governmental, non-governmental, indigenous, and sectoral) has increased in recent years, which has in turn increased the contested nature of the border itself. This observation is in line with the trend toward integrated management of transboundary water resources. It has become more difficult to control phenomena that make the border porous because there are more actors with competing interests. This complexity represents yet another challenge to the border's ontological status. The "piece of real estate" referred to by the anonymous state department official whose cable was published by WikiLeaks is not so simple as it first appears.

Thus the porousness of the border draws its ontological status further into question. On the one hand, the border would not exist without state force to back it up. On the other, states must let go of some of their authority in order to manage their transboundary natural resources effectively. The border as a taken-for-granted institutional line faces challenges from both sides. What, then, of the native point of view?

THE NATIVE BORDER

The native border serves as the third part of the triad that divides this book both figuratively and symbolically. Like mediated and political, the term *native* has more than one valence. First, native means "as seen by indigenous peoples, First Nations communities, or native Indian peoples."[4] The term also evokes "originary," or, loosely interpreted, "existing of its own accord," a meaning that raises the question, "Can borders be native or intrinsic to a place, or must they be imposed by some external force?" Or, to frame the question as Newman and Paasi (1998, 191) might, how has the contested history of the border shaped the construction of native "sociospatial identities"?

This question resonates in recent popular media, such as the 2008 film *Frozen River* about two single mothers, one native and one non-native, who are drawn into the illicit world of cross-border smuggling. The film focuses significantly on issues of borders, border-crossers, and the vague and even frightening environment that can result when traditional tribal and imposed governmental frontier lines collide. In the Great Plains and Prairies, a similarly vague borderland exists at the International Peace Garden straddling the North Dakota-Manitoba border. At a 2011 Métis Festival in the Peace Garden, for example, organizers informed participants, both online and in print, that passports were not required to attend. Although it was true that passports were not required to *enter* the park, immediately upon exiting, participants found themselves at border stations with full complements of guards and dogs where they were required to produce official identification to return from the interstitial "no man's land," where "peace" ostensibly reigns, to their respective countries. For tribal members from the Turtle Mountain Band of Chippewa, whose lands traditionally extended across the border, using tribal identity cards officially issued by tribal governments can be problematic, especially when venturing outside of reservations or attempting to move internationally.

Zalfa Feghali describes a similar situation in her chapter, which opens this section. In July 2010, a group of lacrosse players from the Iroquois Nation attempted to fly to Britain using tribal passports, but "since Iroquois passports are not currently recognized as state-issued travel documents ... the United Kingdom clarified that it would waive the visa requirement and accept the Iroquois document if it was 'accompanied by a United States passport'" (156). This controversy "eventually led the US State Department to confirm in a statement that it had no treaty obligation with the Iroquois nation to recognize their passports, despite the Iroquois 'status' as a nation" (156). Feghali goes on to describe how controversies such as these threaten the status of tribal nations, and mentions how even citizenship in these cases may be a matter of semantics and appeasement as opposed to legitimate realities of official status.

Feghali's broader intervention happens at the level of theory. She investigates Edward Said's (2000) claims about "travelling theory" and the potential that a theory's conceptual or geographic displacement has to animate the questions it poses in new and trenchant ways. In this, Feghali begins with an examination of Mexico-US border in an attempt to challenge the idea that it is "*the* representative border from which border studies as a whole has risen" (154). Feghali focuses on Gloria Anzaldúa, whose writings address gender, sexuality, race, and conditions of (post-)coloniality as they intertwine at the Mexico-US border. Anzaldúa (2007, 25) uses these categories to carve out a space for a border-identity as a place of resistance against the colonial condition, characterized by racism, sexism, and the *herida abierta* – open wound – described earlier. As Feghali looks at how notions of *border-crosser* and *border(lands)* appear in Anzaldúa, she investigates whether these terms and ideas can be mobilized to address issues beyond that border and throughout North America. In particular, she transposes the realities of the Mexico-US border to native communities that straddle the Canada-US border.

One of the central points of reference in Feghali's chapter (as well as Phil Bellfy's later on) is the 1794 Treaty of London, also known as the Jay Treaty, named after British chief negotiator John Jay. It was signed by the United States and Britain at a moment when Britain was at war with France and fervently desired to keep the United States as its ally. The United States wanted Britain to vacate forts that it maintained in the Great Lakes region, and particularly wanted Britain to stop inciting native peoples against the colonies in the Ohio River valley. Although the treaty's first articles are fairly broad and are primarily concerned with tariffs, the circulation of goods, and modes of transportation, later articles contain text that has become something of a focal point as regards tribal claims about free passage across the Canada-US border. Of particular note, the Jay Treaty contains language regarding the movement of people:

> It is agreed that it shall at all times be free to His Majesty's subjects, and to the citizens of the United States, and also to the Indians dwelling on either side of the said boundary line, freely to pass and repass by land or inland navigation, into the respective territories and countries of the two parties, on the continent of America, (the country within the limits of the Hudson's Bay Company only excepted) and to navigate all the lakes, rivers and waters thereof, and freely to carry on trade and commerce with each other. (article 3)

Ironically, although the Jay Treaty accorded rights to Indians "freely to pass and repass by land or inland navigation," post-9/11 changes to border security (such as the Real ID Act) actually hamper border crossings and infringe on rights guaranteed by that treaty.

Feghali provides a partial list of other treaties waived in their entirety by former Secretary of Homeland Security Michael Chertoff, including the Historic Preservation Act, the Endangered Species Act, and the National Environmental Policy Act, in order to extend triple fencing through a Tijuana River Research Reserve near San Diego (159). The fact that few to none of these decisions are subject to judicial review, and that they "affect First Peoples whose tribal lands legally extend across the Mexico-US border" (159), shows the limitations placed on free passage on the southern border of the United States, much like on its northern border.

One way that Feghali proposes an examination of the paradoxes surrounding the legal approval/denial of indigenous passage is through a consideration of the figure of the trickster in Great Plains literature: "I take Gerald Vizenor's trickster analysis as my cue and ... advocate for what Vizenor calls indigenous theorizing" (164). She continues, citing Helen Lock: "Trickster discourse is the process whereby language negotiates the boundaries of the cross-blood's world, deconstructing the fixed, authoritative beliefs and definitions that Vizenor has called 'terminal creeds'" (164).

Joshua Miner makes a similar intervention with respect to notions of the native border. He begins with his observation that although "the political border dividing the native peoples of Canada and the United States is, 'after all, a figment of someone else's imagination,' it certainly bears real-world sociopolitical, cultural, and psychic consequences" (171). Miner contrasts the political border with the native border, and writes, "Native borders are therefore not colonial, because they proceed from nature and defy territorial logic by accounting for natural change and movement" (180). The entire concept of the borderland exceeds itself – the border in Miner's essay serves as a "logic of ... exclusion," where the "Eurowestern eye" watched the "Wild Indian" who resisted progress, with the goal of taming or removing these "savages" by the demarcation of traditional territory into artificial, interstitial lines (171).

Even the term *sovereignty*, signifying self-governance, is called into question here. One of the first people to discuss the concept, Jean Bodin (1993, 74), described it as creating "absolute and perpetual power ... in other words, the most powerful force to rule" (our translation). The term itself, created by a member of the *haute bourgeoisie* and intended to protect against Papal control in the years following the Protestant Reformation, is deconstructed by

Miner as a fragment of a non-native worldview that bears "little resemblance to the principles of Native American societies" (Elvira Pulitano, quoted in Miner, 178).

In this sense of the border as symbol of sovereign control, the border becomes nothing more than an illusion of a false history, a "false pretender ... constructed on a dissimilitude, implying a perversion, a fundamental misdirection" (Deleuze 1969, 295–6, our translation). In this way, Miner provides the first glimmerings of an answer to Feghali's question about what might occur when we begin to take "indigenous theorizing" seriously in order to frame these issues. He describes three Plains writers – Gerald Vizenor from Minnesota's White Earth Band, Louise Erdrich from North Dakota's Turtle Mountain Band, and Blackfoot-Gros Ventre author James Welch – who re-imagine the borderland as a "site of sacred trespass that transforms the body, soul, and language" (173). Border crossing in this sense becomes a form of resistance in and of itself: "Native people have turned colonial abstractions to their advantage, often through movement. Along the Canada-US border, where foreign narratives and bureaucracies have construed them asymmetrically, moving between national spaces becomes a strategy of resistance by evasion" (176). What makes the resistance effective is the way that it challenges Western conceptual pairs such as man/woman, citizen/savage, and human/non-human. If the binary logic that underpins these pairs ceases to hold, it becomes all the more difficult to insist on the us/them distinction made by national boundaries. Miner, borrowing from Vizenor, calls this an "erotic conversion" (173) and traces its effects in the work of the three authors he examines.

Phil Bellfy provides an even clearer answer to Feghali's question about native theorizing and, like Miner, does so by examining the fundamentally undecidable nature of the border. He focuses on the history of the mouth of Lake Superior and draws into question the international dividing line that passes through that great lake. In tone, syntax, and usage, Bellfy's writing can be seen as a manifestation of a challenge to traditional paradigms of control. His style, based significantly on an oral "telling story" dynamic, functions as an unuttered protest to the role of the academy as it props up the very existence of the border he deconstructs. The border, as Bellfy conceptualizes it, came into being without the participation of the Ojibwe/Chippewa, and the discursive acts that established it were directed by non-native European colonial powers at their political rivals, entities that had no a priori jurisdiction over native lands. Subsequent treaties excluded the Ojibwe/Chippewa in similar ways, which further challenged the border's legitimacy for those groups. Even as some treaties created exclusion, others, in particular the Jay Treaty, recognized Ojibwe/Chippewa rights of passage across the border, which created confusion and questioned the inviolate nature of the border and with it the border's legal foundation. Bellfy demonstrates that in addition to the socio-political realities of the legal treaties involved in crafting the reality of the border, the unpredictable vicissitudes of weather and geography, not to mention surveyors' sloppiness and aleatory historical decisions made about various boundaries, challenge even its geography.

Bellfy's essay takes the concept of the border full circle, and as he deconstructs the "common sense" of the hegemonic structures of the border, the discursive (re)production of the border takes on new meaning, which in turn questions the analysis of the border delineated in the first section. The example of border enforcement mentioned in the mediated border section certainly looks different if the border is merely an illusion of history – it is no longer a question simply of enforcing security at the border, but instead of enforcing the ontological status of the border itself.

In this way, Bellfy's challenge to the border's very existence asks us to think beyond the border – not only geographically, but also conceptually. It asks us to think about the nature of the tensions shaping this paradoxical line. We will return to this notion of *beyond* in the book's conclusion.

NOTES

1 For example, the authors of each chapter address the themes listed by Newman and Paasi (1998, 191), suggesting an arrangement along these lines:
- *The suggested disappearance of boundaries*: Michelle Morris's and Paul Sando's essays about riparian rights and river management.
- *The role of boundaries in the construction of sociospatial identities*: Zalfa Feghali's and Phil Bellfy's essays on native identities at the border.
- *Boundary narratives and discourse*: Serra Tinic's and Christopher Cwynar's essays about televisual representations of the border and Joshua Miner's essay about the trickster in native literature.

- *The different spatial scales of boundary construction*: Paul Moore's and Brandon Dimmel's essays on the tensions between local and national identities.

The advantage of the arrangement we have chosen is that it reveals the insights provided by different disciplines into not just the border, but related objects of inquiry, too.

2 On the US side, the Souris River is officially called the Mouse River. The shift from French to English (*souris* is French for "mouse") is yet another signal that a border has been crossed.
3 Recently, scholars of Canadian media have begun to investigate notions of Canadian identity not by looking at the Canada-US relationship but by looking at Canadian production in a global context (e.g., Tinic 2005) or at the range of contradictory forces operating domestically within Canada (e.g., Beaty and Sullivan 2006; Conway 2011).
4 We use the term *Indian* here conscious of its complex implications within post-colonial and indigenous perspectives. For some, it is a pejorative misnomer that perpetuates misconceptions about First Peoples communities and recalls the hubris of early colonists. It is at the same time a term that other Native American writers and communities have embraced as a term of empowerment and pride, as Joshua Miner partially demonstrates.

REFERENCES

Althusser, Louis. 1971. "Ideology and Ideological State Apparatuses (Notes towards an Investigation)." In *Lenin and Philosophy and Other Essays*, translated by Ben Brewster, 127–86. New York: Monthly Review Press.

Anzaldúa, Gloria. 2007. *Borderlands/La Frontera: The New Mestiza*. Third edition. San Francisco: Aunt Lute.

Beaty, Bart, and Rebecca Sullivan. 2006. *Canadian Television Today*. Calgary: University of Calgary Press.

Blake, Gerald. 2005. "Boundary Permeability in Perspective." In *Holding the Line: Borders in a Global World*, edited by Heather N. Nicol and Ian Townsend-Gault, 15–25. Vancouver: UBC Press.

Bodin, Jean. 1993. *Les Six Livres de la République. Un Abrégé du Texte de L'Édition de Paris de 1583*. Chicoutimi: Université du Québec à Chicoutimi. http://classiques.uqac.ca/classiques/bodin_jean/six_livres_republique/bodin_six_livres_republique.pdf.

Brunet-Jailly, Emmanuel, and Bruno Dupeyron. 2007. "Introduction: Borders, Borderlands, and Porosity." In *Borderlands: Comparing Border*

Security in North America and Europe, edited by Emmanuel Brunet-Jailly, 1–18. Ottawa: University of Ottawa Press.

Bukowczyk, John J., Nora Faires, David R. Smith, and Randy William Widdis. 2005. *Permeable Border: The Great Lakes Basin as Transnational Region, 1650–1990*. Pittsburgh: University of Pittsburgh Press.

Conway, Kyle. 2011. *Everyone Says No: Public Service Broadcasting and the Failure of Translation*. Montreal: McGill-Queen's University Press.

Deleuze, Gilles. 1969. *Logique du sens*. Paris: Minuit.

de Loë, Rob. 2009. *Sharing the Waters of the Red River Basing: A Review of Options for Transboundary Water Governance*. Report prepared for the International Red River Board/International Joint Commission. Guelph, ON: Rob de Loë Consulting.

Dombrowsky, Ines. 2007. *Conflict, Cooperation and Institutions in International Water Management*. Northampton, MA: Edward Elgar.

Evans, Sterling, ed. 2006. *The Borderlands of the American and Canadian Wests: Essays on Regional History of the Forty-Ninth Parallel*. Lincoln: University of Nebraska Press.

Habermas, Jürgen. 2001. "The Postnational Constellation and the Future of Democracy." In *The Postnational Constellation: Political Essays*, translated and edited by Max Pensky, 58–112. Cambridge, MA: MIT Press.

Hele, Karl S., ed. 2008. *Lines Drawn upon the Water: First Nations and the Great Lakes Borders and Borderlands*. Waterloo, ON: Wilfrid Laurier Press.

Horkheimer, Max, and Theodor W. Adorno. 1972. "The Culture Industry: Enlightenment as Mass Deception." In *Dialectic of Enlightenment*, translated by John Cumming, 120–67. New York: Continuum.

Isern, Thomas D., and R. Bruce Shepard. 2006. "Duty-Free: An Introduction to the Practice of Regional History along the Forty-Ninth Parallel." In *The Borderlands of the American and Canadian Wests*, edited by Sterling Evans, xxvii–xxxv. Lincoln: University of Nebraska Press.

Jameson, Elizabeth, and Sheila McManus, eds. 2008. *One Step over the Line: Toward a History of Women in the North American Wests*. Edmonton: University of Alberta Press.

Konrad, Victory, and Heather N. Nicol. 2008. *Beyond Walls: Re-inventing the Canada-United States Border*. Burlington, VT: Ashgate.

Luebke, Frederick C. 1984. "Regionalism and the Great Plains: Problems of Concept and Method." *Western Historical Quarterly* 15: 19–38.

Miller, Mary Jane. 1993. "Inflecting the Formula: The First Seasons of *Street Legal* and *LA Law*." In *The Beaver Bites Back? American Popular Culture in Canada*, edited by David. H. Flaherty and Frank E. Manning, 104–22. Montreal: McGill-Queen's University Press.

Newman, David, and Anssi Paasi. 1998. "Fences and Neighbours in the Postmodern World: Boundary Narratives in Political Geography." *Progress in Human Geography* 22 (2): 186–207.

Rutherford, Paul. 1993. "Made in America: The Problem of Mass Culture in Canada." In *The Beaver Bites Back? American Popular Culture in Canada*, edited by David H. Flaherty and Frank E. Manning, 260–80. Montreal: McGill-Queen's University Press.

Said, Edward. 2000. "Traveling Theory." In *The Edward Said Reader*, edited by Moustafa Bayoumi and Andrew Rubin, 195–217. New York: Vintage.

Smythe, Dallas. 1981. *Dependency Road: Communication, Capitalism, and Consciousness*. Norwood: Ablex.

Taylor, Charles. 1993. *Reconciling the Solitudes: Essays on Canadian Federalism and Nationalism*. Montreal: McGill-Queen's University Press.

Tinic, Serra. 2005. *On Location: Canada's Television Industry in a Global Market*. Toronto: University of Toronto Press.

United States. 2008. Department of State, US Embassy in Ottawa. *Subject: Primetime Images of US-Canada Border Paint US in Increasingly Negative Light*. 25 January. Published by WikiLeaks 1 December 2010. http://www.wikileaks.ch/cable/2008/01/08OTTAWA136.html.

THE MEDIATED BORDER

1

The Borders of Cultural Difference: Canadian Television and Cultural Identity

SERRA TINIC

The saliency of the border as both a physical, cartographical point of demarcation as well as a marker of cultural similarity and difference cannot be overstated – at least on the Canadian side of the continental dividing line. The United States has always loomed large in the Canadian imagination. Sitting next to one of the world's cultural and political superpowers has resulted in what many political sociologists refer to as a sense of *negative identity*, or the propensity to define yourself by what you are not. It is a particularly acute phenomenon when national neighbours of unequal power status share a language and similar economic, cultural, and political systems. Consequently, for many Anglo-Canadians, national self-definition has long manifested itself as "not being American."

In this essay, I focus on the central role that media, and specifically television dramas and comedies, have played in debates over cultural sovereignty and the maintenance of a border of distinction between the two countries. Indeed, television has arguably been the central cultural industry through which Canadian policymakers have sought to foster a sense of national community through a discourse of cultural defence against Hollywood's cross-border influence. My own past field research departs from the Fortress Canada arguments that defined early broadcasting debates and instead explores the contradictions in the Canada-US media relationship. Here, the proximity to the United States can be seen to provide opportunities as well as constraints for domestic television producers. By painting a picture with very broad brush strokes, I examine the range of border crossings, both domestic and international, that inform the televisual representations and production partnerships that mark the cultural experiences of space and place in Canada today. I conclude with a brief consideration of a new socio-cultural border that is emerging between Canada and the United States in post-9/11 television narratives. As recent excerpts from the controversial WikiLeaks documents reveal, US State Department officials have been watching Canadian television with renewed interest and great consternation. The geopolitical tensions between the two nations over the Iraq war, and the somewhat more amorphous war on terror, have transformed the "world's largest undefended border," both figuratively and literally, into the "world's largest secure border." In contemporary Canadian television dramas, the forty-ninth parallel is now a dominant trope of cultural difference and national self-confidence that appears to have taken both Canadians and Americans by surprise.

But first a bit of broadcasting history is in order. The late communications scholar Dallas Smythe (1981) argued decades ago that if Canadian policymakers had been serious about fostering a continentally independent broadcasting system, they would have adopted a broadcast standard different from that of the American NTSC model. This would have effectively created a transmission barrier to American over-the-air signals and, therefore, cultural influence. The failure to do so, in his opinion, resulted in a situation whereby Canada became a cultural colony or branch plant of the American media industries. I do not personally subscribe to this isolationist prescription as it illustrates an adherence to an outdated cultural imperialism thesis. It is, however, difficult to argue for a more effective method of denying the power of the Hollywood juggernaut than to batten down the hatches and close the virtual border to the influence of American popular culture. However, denying Canadians the opportunity to engage with US entertainment was never the intent behind national media development policies. In fact, I imagine such a proposition would have been met with outrage as most Canadians living close to the border had already become avid fans of American stories in the early days of radio border spillover. Rather, policymakers aimed to guarantee a space for the production and circulation of culturally specific stories so that Canadians could imagine themselves as a nation living within different circumstances than those of their more dominant southern neighbour.

Consequently, national public broadcasting was developed according to the loftiest principles with unfortunately short-sighted practical strategies. The Aird Commission, which established the Canadian Radio Broadcasting Commission (CRBC), the 1930s forerunner to the Canadian Broadcasting Corporation (CBC), believed a national public radio system was so vital to the evolution of the nation-state that the government rapidly built one of the most sophisticated technological systems to connect far-flung regional communities across the expanse of one of the largest countries in the world. Unfortunately, the ensuing expense meant there was little money left over for the production of actual content. The end result was that in the early years of the CRBC, Canadians experienced everything the Aird Commission had hoped to resist, which was the centralization of production in Ontario as well as the actual import of some American programming. These problems have plagued every successive task force and royal commission on broadcasting. Trying to find a balance between economic contingencies and domestic cultural expression became what Marc Raboy (1990) has referred to as a series of "missed opportunities" in Canada – basically the lack of financial and political will to fully invest in a truly publicly funded domestic media industry. This manifested itself not only in the establishment of competing private and public networks but also within the structure of the CBC itself, which has relied on advertising for one-third of its production revenues for the last sixty years.

My rationale for providing this rather superficial history of Canadian policy is to underline the continuing saliency of the tension between building an industry ("the market or economic" goals) and fostering a sense of cultural community through television narratives. Today, a new generation of domestic television producers and distributors have adapted to this industry-culture schism by moving across local, national, and transnational networks simultaneously. In doing so, they have both put themselves at the forefront of new global production practices as well as questioned the primacy of the nation as a stable or essentialized site of cultural representation.

Canadian television production has always been informed by competing cultural tensions at all levels: the *domestic*, read as regional alienation versus the nation building goals of policy; *continentalism* as both the allure of cultural proximity to the United

States and a simultaneous defensive emphasis on cultural distinction; as well as a gaze toward the *global* both through the diverse communities within the country but also out of perceived economic necessity. The ways in which Canadian television producers have manoeuvred through these structural constraints has put them at the forefront of many of the issues defining today's media industries transnationally, specifically in regard to the balancing act between culturally specific expression and the homogenizing potential of global audience market expansion. It is a truism that television drama is one of the most expensive genres to produce and in Canada, with such a small domestic market, the ability to achieve the requisite production quality to attract and maintain sizeable audiences has depended on a combination of government subsidies and an eye to international sales. In an era of ever-decreasing support for national public broadcasting, Canadian producers have long relied on international partnerships in the form of official co-productions and joint ventures to mitigate against economic uncertainty. With over fifty official treaties, Canada is the global leader in international joint ventures and its success in this sector can be largely attributed to the country's cultural and geographical relationship with the United States. Canadian producers are attractive partners as they understand the grammar of American television. They are viewed as cultural intermediaries who can speak the language of Hollywood but are still seen as having a slight European sensibility that allows for nuanced differences in televisual storytelling (Armstrong 1996). Often, Canadian producers and their co-production partners do seek to erase culturally specific markers and emphasize American formulas, genres, and settings to increase their international sales value. *Outer Limits* and *Stargate* are but two examples of the types of generic or industry productions that are often seen as hallmarks of the culturally homogenizing impact of capital interests in a global media environment, to say nothing of Hollywood's continuing dominance transnationally. If we combine these types of co-productions with the fact that Canada has also served as the largest site of American television production outside of New York or Los Angeles, then it would be easy to conclude that Canada's global role in television production is to serve either as a branch plant for Hollywood or as a mimetic voice in formulaic storytelling. However, this type of macro-level analysis assumes that Canadian producers dwell in an either/or world – looking solely to global sales or the more romanticized notion of looking only inward to somehow express that elusive but somehow unique Canadian sensibility in a television series. In my own research I have found that most creative personnel move between industry and service productions to more locally inflected works as necessity and desire dictate. Furthermore, this flow between production worlds has strategically positioned them to exploit the rapid changes at work within the global television industries, including that of the United States (Tinic 2005).

More than ten years ago, when I first began fieldwork in Vancouver, or Hollywood North as it is often called, it would have been an understatement to claim that domestic drama was in a state of crisis in Canada. At the time, there were very few television series in production, and virtually no English Canadian dramas were being produced outside of Toronto. This was due to both the devastating cutbacks to the CBC that began in the late 1980s as well as the network's belief that regional producers were somehow incapable of producing quality television that spoke to a national audience. As a result, many independent production houses took advantage of the American presence in Vancouver and fully immersed themselves in service work for Hollywood runaway productions or in global joint ventures with other countries. Several writers, directors, and producers even began dual careers, so to speak, by going to Los Angeles to work on American television series while simultaneously developing and show-running their own productions back in Vancouver. For these individuals, the turn away from the national was not driven by a desire to move into globally generic production with the hopes of achieving American success. Rather, they saw the emergent cross-border production relationships as a means to accrue the skills and financial resources (and all-important clout) that would allow them to develop the types of locally-inflected programs that they felt were under-supported by national regulators and funding agencies. They saw going global as a means to eventual local success.

By the mid-1990s, one of a long list of broadcasting task force committees submitted a federal report that stated the CBC would have to radically increase its level of specifically domestic television content if it was to remain a relevant institution in the new millennium (Mandate Review Committee 1996). Thus facing a dearth of television drama within a restructured network that would be expected to move toward out-of-house production, CBC administrators were forced to suspend their disbelief that regional producers could contribute nationally resonant programming. As scripts and story ideas flooded into the network, the Vancouver production centre was best positioned to contribute deliverable content because of the momentum that the American locations industry there had generated and sustained. The CBC picked up one series, the crime procedural *Da Vinci's Inquest*, out of approximately six finalists. *Da Vinci* was premised on the real-life experiences of Vancouver's crusading coroner and later mayor, Larry Campbell, and the CBC promoted it as a cross between the American shows *Homicide: Life on the Street* and *Quincy*. Much was made of the fact that the show's creator Chris Haddock had written for American runaway television productions, most notably *MacGyver*, and had established connections with American studios. The series was given the green light because it combined a proven producer with a popular genre. CBC executives were confident that it had international legs and would be globally marketable (Ontario Scene 1997). These types of statements fuelled a great deal of pessimism amongst the Vancouver production community because all signs pointed to a move toward that homogenizing model of marketable television in the Hollywood mould. Many feared it represented precisely the type of production colonization or branch plant mentality that Smythe discusses. However, *Da Vinci* not only quelled all concerns, but also was truly a landmark moment for Canadian television. The series was explicitly rooted in the socio-cultural experiences of the local community. All of the episodes drew on the consequences of poverty, prostitution, and IV drug use that marked the impact of the global capital transformation of the Downtown Eastside, a four-block radius in Vancouver that is the poorest and most at risk neighbourhood in the nation. In fact, in 1997, this one neighbourhood was found to have the highest rate of HIV infection in the industrialized world according to the World Health Organization (Riley 1998). The series was ambiguous, probed social issues, and was dark and grim in the best risk-taking sense of public broadcasting. It was also incredibly popular. For ten years, *Da Vinci's Inquest* was the highest rated domestic drama in Canada. It was also quickly sold to forty-five countries and, to much national pride, was the first openly Canadian series to be purchased by an American broadcast network – CBS, which desperately needed to fill the void of syndicated scripted programming that had resulted from the overproduction of reality television.

Da Vinci represents a turning point in both Canadian domestic and global television goals. The series emerged out of a moment of crisis for the CBC and, because of Chris Haddock's familiarity with Hollywood, opened the door to one of the world's most restricted markets, the American broadcast networks that were facing their own content crisis of the moment. The fact that the series received critical acclaim and high ratings in the United States exposed the fallacies underlying the conventional logic that had come to define decision-making structures in Canadian network circles. First, the success of the series showed that character-driven drama from an explicitly regional perspective could resonate nationally and, perhaps more significantly, that global audiences (and Americans in particular) could appreciate television drama that didn't attempt to hide its point of national origin. In this respect, *Da Vinci* set the stage for what would become a much more confident national production infrastructure in ways that resist theories of cultural homogenization, American cultural imperialism, and the essentialization of what constitutes the national in global media representations.

POST-9/11 TELEVISION IN CANADA

Being attached to America these days is like being in a pen with a wounded bull. Between the pot smoking and the gay marriage, quite frankly it's a wonder there is not a giant deck of cards out there with all our faces on it.

Rick Mercer, quoted in the *New York Times*

In 2003, the *New York Times* (Krauss 2003) invoked the above segment from the popular Canadian comedian Rick Mercer's satirical news comedy, the *Rick Mercer Report*, as part of a larger analysis of the increasing cultural and political tensions between Canada and the United States. Probably much to the astonishment of most Canadians, the *New York Times* declared that the Canadian identity crisis vis-à-vis the United States had finally been resolved in the years following the attacks on the World Trade Centre on 11 September 2001. According to the article, Canada was becoming more "European" in its sensibilities and the once "most-favoured" definition of the relationship between the two countries was becoming marked by discord and tension. Indeed, the Canadian government's popular decision not to follow the United States to war in Iraq did appear to underline a new sense of independence and self-confidence north of the border. As American politics and social discourse grew increasingly conservative, the Canadian propensity for a negative sense of national identity waned, and the continental drift between the two nations was increasingly evident in the nightly news coverage of American politics. The emergent sense of Canadian cultural difference became a prominent theme in domestic television drama and comedy production.

Two Canadian television series, *Little Mosque on the Prairie* (2007–12) and *The Border* (2008–10), explicitly drew on the post-9/11 cultural milieu in their story arcs and underlined the ideological divisions between the two countries. The comedy *Little Mosque on the Prairie* is based on the series' creator Zarqa Nawaz's observations of life as a Muslim in North America and, more recently, her cultural experiences as a resident of the prairie city Regina, Saskatchewan. Although the series draws most of its humour from the cultural miscommunications that occur amongst the inhabitants of a fictional small prairie town, story-lines frequently play on the increasing sense of Islamophobia that seems to prevail in the United States. Consequently, the series' narratives often invoke no-fly lists and unyielding customs officers. The drama *The Border* is even more forthright in its depiction of the changing political landscape in North America. Indeed, the series is premised on the fact that what was once considered the world's largest undefended border has now become a trenchantly policed and guarded dividing line. *The Border*'s central characters are officers with the immigration and customs security agency who are forced into a partnership with a homeland security agent who personifies the political and cultural differences between the two countries' approaches to policing specifically and governance in general. The central narrative of the series focuses on the Canadian agents' consistent "education" of the archetypical American militant officer in matters of tolerance to cultural difference and respect for jurisdictional differences in matters that range from native land claims to attitudes about terrorism.

Although both *The Border* and *Little Mosque* were initially popular in Canada, neither series was able to sustain a sizeable domestic audience in comparison to American television imports. Consequently, it became national news when the 2010 WikiLeaks documents were released and revealed that the US State Department had been following both series intently and was seriously concerned with their perceived impact on diplomatic relations between the two countries:

> The Canadian Broadcasting Corporation (CBC) has long gone to great pains to highlight the distinction between Americans and Canadians in its programming, generally at our expense. However, the level of anti-American melodrama has been given a huge boost in the current television season as a number of programs offer Canadian viewers their fill of nefarious American officials carrying out equally nefarious deeds in Canada ... Canadian broadcast entities, including those financed by Canadian tax dollars, twist current events to feed long-standing negative images of the United States. (WikiLeaks Archive 2010)

To equate "distinctions between Americans and Canadians" with "anti-Americanism" marks a lack of understanding of the sense of negative identity that long dominated Canadian cultural discourses. Namely, that the American cultural other permeates the broader Canadian cultural imagination and that domestic broadcasting policy has long stressed that a televisual border of difference needed to be fostered if Canadians were to see themselves as a unique imagined community. Ironically, the state department's alarm is evidence that, in recent years, the CBC has fulfilled its mandate.

REFERENCES

Armstrong, M.E. 1996. "TIFF Buyers Focus on Ancillary Markets." *Playback*, 9 September, sec. 9, 3.

Krauss, C. 2003. "Canada's View on Social Issues is Opening Rifts with the US." *New York Times*, 2 October. http://www.nytimes.com/2003/12/02/world/canada-s-view-on-social-issues-is-opening-rifts-with-the-us.html.

Mandate Review Committee, CBC, NFB, Telefilm. 1996. *Making Our Voice Heard: Canadian Broadcasting and Film for the 21st Century*. Ottawa: Department of Canadian Heritage.

Ontario Scene. 1997. *Playback*, 23 September, 14.

Raboy, M. 1990. *Missed Opportunities: The Story of Canada's Broadcasting Policy*. Montreal: McGill-Queen's University Press.
Riley, D. 1998. *Drugs and Drug Policy in Canada: A Brief Review and Commentary*. http://www.parl.gc.ca/Content/SEN/Committee/371/ille/library/riley-e.htm.
Smythe, D. 1981. *Dependency Road: Communication, Capitalism, and Consciousness*. Norwood: Ablex.
Tinic, S. 2005. *On Location: Canada's Television Industry in a Global Market*. Toronto: University of Toronto Press.
WikiLeaks Archive 2010. http://www.nytimes.com/interactive/2010/11/28/world/2010112.

2

The Canadian Sitcom and the Fantasy of National Difference: *Little Mosque on the Prairie* and English-Canadian Identity

CHRISTOPHER CWYNAR

INTRODUCTION

Serra Tinic's essay in this collection provides an ideal point of entry to a discussion of the Canada-US border as an arbitrary line that divides on a cultural level, or rather, as a line that prompts active divisions on a cultural level. Tinic provides a brief recap of Canadian broadcasting history that focuses on the perceived threat of American economic and cultural imperialism resulting from a combination of geographical proximity, cultural and linguistic similarity, and the power imbalances between the mighty American and limited Canadian cultural industries. Of course, the airwaves observe no borders and the persistence of the American media industry has had an undeniable impact on its sparsely populated northern neighbour. Canada, as Harold Innis famously observed, has been twice colonized (cited in Berland 1996, 58). In English Canada, the enduring links to Great Britain – the dominant imperial presence of a previous age – make the terrain susceptible to US cultural imperialism and simultaneously provide cultural resources with which to resist it.[1] In this context, the trajectory of Canadian television has been defined by a tension between those who see the need to protect Canada from American economic and cultural imperialism and those who wish to consume American content. It is thus no surprise that the contemporary English-Canadian telescape largely features a strange blend of American content, British content, inexpensive quota-satisfying programming produced in Canada, and overtly nationalistic content.[2] In this context, national identification often predicates on the contrast between content that registers as "American" and that which registers either as "not American" or is coded as distinctly "Canadian."

Yet these elements cannot always be so easily separated. In fact, nationalist discourses may obscure complex cultural processes in which two or more sides work together to produce ambivalent texts that combine codes, conventions, and cultural references that allude to external influences along with elements that invoke a specifically Canadian experience. As Frank Manning (1993, 3) argues, "Popular culture crosses the border, but it is also perhaps the most prominent marker of what that border, an 'open' but highly symbolic boundary, can mean to those for whom its presence is a proximate and pervasive reality." In other words, the admittedly uneven flow of texts, genres, and discourses across the border provides abundant opportunities for the actualization of the divide between the two countries.

In such instances, the Canadian element of a given experience is often located in the processes of negotiation that occur on the production or usage (often termed "consumption") side. Manning (1993, 3) argues for the centrality of this practice to Canadian cultural life, and suggests that Canadians have a "peculiar ambivalence" to American popular culture that results from the pervasiveness of US popular culture in Canada and the extent to which this material influences the production of *Canadian* popular culture. Mary Jane Miller (1993) and Aniko Bodroghkozy (2002) each address this practice with respect to Canadian television.[3] Miller's inflection theory addresses the ways in which certain Canadian programs subtly alter styles and genre formulas defined in other contexts in order to produce local statements. Bodroghkozy builds on the work of Manning and others through her application of active audience theory to Canadian television in order to conceptualize the ways in which Canadian cultural experiences might emerge out of engagement with cultural material that is coded as American. Her analysis of *Street Legal* (CBC, 1987–94) and *Due South* (CTV, 1994–96) acknowledges the political-economic factors at play in Canadian television production and usage, but goes beyond those conditions to examine the forms of negotiation and engagement that these local (or localized) products might facilitate for Canadian audiences.

Little Mosque on the Prairie (CBC, 2007–12) appears to function in a similar manner. Tinic notes that the program, along with *The Border* (CBC, 2008–10), provides ample opportunity for negative identification in the manner in which it draws on "the post-9/11 cultural milieu in their story arcs and underlined the ideological divisions between the two countries" (36). Tinic observes that, although the series derives most of its humour from the cultural miscommunications that occur among the inhabitants of a fictional small prairie town, these miscommunications draw on the increasing sense of Islamophobia that seems to predominate in the United States. The implication here is that this "Islamophobia" is not characteristically Canadian (though there is a significant amount of anti-Muslim sentiment in Canada). This interpretation is first based on the implicit contrast between the ostensibly "multicultural" Canadian nation and the American "melting pot" that cannot manage to accommodate the Muslim "other." It also refers to the ways in which viewers might interpret the program's depictions of those who would propagate and internalize negative stereotypes of Muslims as representing implicitly American elements in this context.

Little Mosque is set in Canada and its cultural politics are further complicated by issues that pertain to genre and intertextuality. On the one hand, *Little Mosque* seems to be entirely consistent with Canadian – and Canadian Broadcasting Corporation (CBC) – television conventions: it is earnest and informative, addresses Canada's particular inter-regional dynamics, and displays a strong current of social liberalism in its emphasis on multiculturalism and concern for issues that relate to gender and sexuality. Yet the program is also a deceptively conventional sitcom that draws extensively on a shared Anglo-American popular cultural heritage dominated by the United States. It reflects an enduring ambivalence in Canadian cultural texts between elements that are explicitly marked as Canadian and those that implicitly or explicitly invoke other places. *Little Mosque* is thus a local statement made through the adaptation of a format defined in other places and then aired on the CBC, an iconic national institution with its own complex set of mythologies.

There is also the matter of *Little Mosque*'s international success and high initial ratings in Canada.[4] The program's degree of visibility nationally and internationally, along with its culturally ambivalent textual properties, make it an ideal object through which to interrogate both the textual representation of "Canadianness" on television and the manner in which a program's network context can frame its consumption in a crowded telescape. This approach is in part a response to recent laments about the dearth of studies that address "the cultural character of Canadian television" (Druick and Kotsopoulos 2008; Urquhart and Wagman 2006, 2).[5] With this in mind, I engage in a textual analysis of the program in conjunction with its network context along with selective references to its critical reception within Canada in order to elucidate the preferred readings that might be activated when this program is viewed in English Canada. I argue that *Little Mosque* embodies this cultural ambivalence through its inflection of genre codes and conventions that have been overdetermined in other contexts and through the ways in which it deploys satire and parody.

In the first section, I examine *Little Mosque* in terms of Canadian multiculturalism, the program's use of interconnected fish-out-of-water situations, and its attempts to inform its viewers about Muslim religion and culture. I discuss these examples as Canadian inflections on the situation comedy form, which the American cultural industries have overdetermined. In the second section, I address the ways in which *Little Mosque*'s satirical elements might function to provide opportunities for negative national identification in relation to the United States. At the same time, I note that the program's use of parody as intertextual shorthand reaffirms the centrality of a broader popular culture heritage that the American cultural industries dominate. In the conclusion, I contend that these elements embody Canada's ambivalent position as a culturally and economically marginal entity within a globally dominant bloc. *Little Mosque* epitomizes the ongoing struggle to negotiate this position, which might be a quintessentially Canadian practice with respect to popular culture.

AN (UN)COMMON COMEDY: *LITTLE MOSQUE* AS A CANADIAN TELEVISION PRODUCT

Little Mosque is produced for the CBC by the Toronto-based companies Westwind Pictures and FUNdamentalist films. Filmed in Saskatchewan and Toronto, the series was created by Zarqa Nawaz, a British-born Canadian Muslim of Pakistani origins. She is a former documentary filmmaker whose films primarily concerned Islam in the West, particularly the role of women in the religion.[6] In her work, Nawaz has frequently sought to write against stereotypical representations of Islam through examinations of Muslim communities in Canada.

With *Little Mosque,* Nawaz shifted to the situation comedy as a vehicle through which to expose and interrogate these representations. The program draws on her experiences as a Toronto resident who moved to small-town Saskatchewan. *Little Mosque* sees a diverse Muslim congregation in the fictional small town of Mercy establish a mosque in the rented basement of the local Anglican church and then struggle to negotiate its position in relation to the broader community. The ensemble cast features characters that espouse a diverse range of viewpoints, from the conservative economics professor and community elder Baber Siddiqui and the no-nonsense cafe owner Fatima Dinssa, to the liberal contractor Yasir Hamoudi and his wife Sarah, who are augmented by their daughter Rayyan and the young imam Amaar Rashid. These two represent the next generation with its ever more progressive approaches to religion and social values. The Christian community is represented by the suspicious and conservative radio host Fred Tupper, the liberal Anglican Reverend Duncan Magee (replaced by the less sympathetic Rev. Thorne in the fourth season), the ignorant but deceptively sweet farmer Joe Peterson, and the scheming mayor Ann Popowicz, whose surname alludes to the waves of Eastern European immigrants that settled the Canadian prairies in the early twentieth century.

Given the nature of this smaller community within the broader population, the Muslim group must practise a form of strategic essentialism in order to protect its interests. At the same time, the differences between members of the community provide for comedic scenarios that interrogate certain Canadian cultural norms, affirm others, and examine intersectionalities that combine to produce individual subjectivities. Confrontations and misunderstandings within the Muslim community, and between that community and the broader Christian community, provide opportunities to expose and consider differences based on religion, gender, age, region, and other discursively produced markers of identity. For example, the young feminist Rayyan consistently clashes with the conservative elder Baber while Fred frequently uses his radio program to cast suspicion on the Muslim congregation.

With the notable exception of Matheson (2012), most of the other scholarly studies of *Little Mosque* tend to consider the program largely in terms of an East-West dichotomy in a Western mass media context and the program's depiction of women and the struggle for gender equality (Cañas 2008; Khan 2009). While this work makes valuable contributions, it cannot account for the medium and nation-specific dynamics inherent in a CBC television production that aired in a US-dominated Canadian telescape in the decade after 9/11. This work effectively ignores the manner in which both Canada's national public broadcaster and the Canadian media have framed the program. However, these factors are crucial to assessing the actual impact of *Little Mosque* on Canadian viewers and those who may read or hear about the program.

The fact that *Little Mosque* is a CBC product that is primarily consumed through the broadcaster's television network and website in Canada is particularly significant. The CBC has long occupied a special place in the Canadian mediascape. Maurice Charland's (2004) "technological nationalism" theory contends that the CBC has been elevated to national icon status within the Canadian mythos by virtue of its perceived ability to unify the country. Charland details the circumstances surrounding the CBC's emergence, including the desire of cultural nationalists to protect Canada's mediascape from putatively debasing American "mass" popular culture. In an attitude that reflects the "Anglophilic nationalism" of the time, the CBC was to present informational news and edifying cultural programming that would uplift the people and produce a nation of citizens rather than consumers (Edwardson 2008, 12).

This initial orientation resulted in an institutional emphasis on news and documentary programming, which developed into something of a Canadian public broadcasting tradition.[7] In the television era, the CBC's orientation towards such programming has been augmented by the production of realist drama and satirical sketch comedy programming as part of an effort to provide for a Canadian presence on television (Hogarth 2002). These efforts speak to the desire of cultural nationalists to write against the dominant forms of television programming. As Bart Beaty and Rebecca Sullivan (2006, 68) note, "Anxiety over Canadian programming is deeply tied to notions of television as an inherently lowbrow medium that could, if suitably linked with good national values, be legitimated in terms of its service to the State, rather than through its aesthetic content." The

CBC has been the primary venue for this project. In a comparison between the CBC and Canadian commercial networks CTV and Global in recent years, Beaty and Sullivan (2006, 70) relate that the latter two devote most of their resources to acquiring American content while the "CBC, on the other hand, with its focus on mostly Canadian television (with the notable exceptions of American movies and British soap operas), seeks to distinguish itself through appeals to traditional Canadian nationalism."

Given these objectives, it is perhaps not surprising that the CBC or its private competitors in Canada have not emphasized the situation comedy. In fact, critics and scholars have accounted for this via economic explanations (sitcoms are costly to produce and often fail, making the US industry better equipped to produce them), service explanations (the Canadian sitcom market has been effectively served by US producers during the mass-media era), and cultural hierarchies (the genre has not been valued in the same way as the others). Geoff Pevere and Greig Dymond's (1996, 204–9) microhistory of the Canadian sitcom chronicles the various failed attempts to generate entrants into the genre. They suggest that the "Canadian sitcom paradox" is the result of a situation in which the best talent departs the country and a paucity of potential programs further increases the chances of inferior programs being produced. Pevere and Dymond note that many of the sitcoms produced in Canada have been highly derivative of successful programs that originated in other contexts. There is also the fact that, despite the popularity of foreign sitcoms among Canadian audiences, critics and scholars have seldom conceived of them as authorized vehicles for national identification or imagination. The implication seems to be that one may watch plenty of sitcoms in Canada, but they are not the sites through which individuals are encouraged to imagine the national community or identify as Canadian subjects (Anderson 1991).[8]

The exception to this may be situations in which the genre can be deployed to work through issues that pertain to cultural difference. It may be true, as Byers (2009) suggests, that English Canada lacks an established tradition of ethnic comedy, but both *King of Kensington* (CBC, 1975–79) and *Little Mosque* have used the sitcom format to interrogate cultural relations in an officially multicultural society (Byers 2009). It is worth noting here that the only truly successful Canadian sitcom prior to the 1990s performed a similar sort of cultural labour. *King* starred Al Waxman as a Jewish grocer in

Toronto's diverse Kensington Market neighbourhood. Sarah A. Matheson (2006, 48) contends that the program had a conciliatory approach that helped the country negotiate the divisions that emerged with Quebec's sovereignty movement.⁹ This idyllic vision of multicultural Toronto embodied the "peaceable kingdom" vision that Trudeau first articulated in the late 1960s. Matheson (2006, 48) writes, "It is around issues of ethnicity and gender that these efforts towards reconciliation typically unfold. In a Canadian context, *King* is an especially useful text through which to examine how television may be used to fashion national images and provide models of national belonging." *Little Mosque* functions in a similar fashion; where *King*'s Kensington Market has been read as a microcosm for Canada, *Little Mosque* uses its "situation" to conduct a humorous examination of the relationship between Muslims and Christians in small-town Canada. In each of these cases, the sitcom provides an ideal vehicle through which to dramatize the negotiation of difference in multicultural Canada, and thereby reaffirms national ideals of tolerance for ethnic and religious diversity.

In fact, Matheson (2012) addresses the ways in which *Little Mosque* functions in this manner and considers them in relation to the critical responses to the program inside and outside of Canada. Matheson finds that the sitcom provides an effective vehicle through which to interrogate stereotypes and examine the clashing of cultures, but that conventions of the genre also limit the scope of these considerations in a reductive manner. Matheson contends that the conventional sitcom depictions of the middle class and the nuclear family dynamic facilitate Nawaz's attempts to present Muslims in terms of normative Western values. The program thus writes against depictions of the Muslim other, but it does so through the use of universals that effectively propagate an idealized image of Canada as a multicultural nation while ignoring the actual lived experience of multiculturalism. Invoking Richard Day, Matheson suggests that a "fantasy of unity" underpins the perception of the program in relation to the idealized conception of Canada as a multicultural nation. The established conventions of the American sitcom help to facilitate the maintenance of this fantasy as they present this conspicuously "different" group in a manner that allows for its easy incorporation into the prevailing images of Canada.

These genre conventions do not originate in Canada, yet they are put to use here in order to develop an idealized image of the Canadian nation. I thus argue that they might also facilitate the production of a fantasy of difference along national lines in terms of the way in which *Little Mosque* weds sitcom convention to the Canadian multicultural metanarrative. The latter might provide Canadian viewers with an opportunity for national identification in relation to the United States. Manning (1993, 7) argues, "As a national ideology and public value system, the significance of multiculturalism derives from its capacity to represent a clear and simple distinction to the alleged monoculturalism of the United States." The emphasis on diversity in a multicultural society also provides an immediate point of distinction for Canadian viewers given that this Canadian mosaic approach is understood to contrast strongly with the American melting pot model (Stratton and Ang 1998).[10] Although numerous scholars have critiqued Canada's multicultural ideal in recent years, there are indications that it still functions as a primary national metanarrative for many Canadians.[11]

The sitcom is useful in this particular context because its conventions provide a means to interrogate difference in a non-threatening manner while they reaffirm the multicultural metanarrative that is so central to the modern Canadian nation. In effect, *Little Mosque* is "inflecting the formula" of a genre that has been defined elsewhere (Miller 1993, 104–22). This term refers to the manner in which Canadian television programs often localize these genres through alterations on thematic, formal, and aesthetic levels in order to create culturally "Canadian" content. *Little Mosque* engages in this through a number of subtle (and not-so-subtle) inflections on the conventions of the sitcom format. These inflections also resonate with the CBC's own mythos and may be amplified by it in the Canadian context.

The most immediately apparent inflection involves the program's title, which tweaks that of the classic American drama series *Little House on the Prairie* in order to signify that it will address the introduction of a foreign element into the Canadian heartland, thereby reconfiguring narratives of settlement and colonization to suit a new scenario (Byers 2007). This alludes to the manner in which the program uses this comedic situation to examine how the Christian and Muslim populations conduct cultural negotiations on an individual and collective basis in a small Saskatchewan town. This central premise resonates with the CBC's mandated objective of reaffirming Canada's multicultural society.[12] As previously discussed, this premise affirms the multicultural ideals that inform the hegemonic conceptions of the Canadian nation-state and provides citizens with a positive basis for negative identification in relation to the perceived assimilation orientation of American society.

Little Mosque deploys two interconnected fish-out-of-water plotlines that overlay regional difference and cultural difference in a manner that has prompted one scholar to characterize it as a "traditional Canadian sitcom" (Byers 2009). The program presents a local variant of the conventional fish-out-of-water premise wherein a diverse group of Muslims must engage in a form of strategic essentialism in order to practise their faith in a small Midwestern community. The cultural negotiations between the Muslims and the wider Christian community at times reveal incommensurable or irreconcilable differences within and between these communities, though the program generally upholds the liberal-pluralist ideals that have predominated during the era of official multiculturalism. In its refusal to merely perpetuate pernicious stereotypes and its interest in meditating on the nature of difference, *Little Mosque* distinguishes itself from the prevailing representations of Muslims and Islam in Western popular culture.

The series' second fish-out-of-water premise concerns Amaar, a former lawyer who moves to Mercy from Toronto to serve as the mosque's imam. This premise creates many opportunities for the program to play with the centre-periphery relationship between the eastern metropolis and the small western town. This taps into the profound divisions between the urban and the rural and between Canada's various regions, which Byers (2009) describes as "'essential' Canadian sitcom fodder."[13] In fact, regionalism has been a defining – and anxiety-producing – feature in Canadian cultural discourse since confederation. The literary critic Northrop Frye (1971, ii) observes that "identity" is "regional, rooted in works of imagination and culture" while "unity" must be "national in reference, international in reference, and rooted in a political feeling." He argues that this results in a "political sense of unity and the imaginative sense of locality" that frequently recurs in the Canadian psyche. As Charland notes, the CBC was created in part to resolve this tension and unify the nation (he contends that it has been celebrated for its ability to do so). Fostering inter-regional communication and understanding remains a core institutional objective, even if the CBC has become progressively more centralized in the decades since its inception (Tinic 2005, 3–28). In this case, the regional centre-west dynamic is a useful national counterpoint to the larger orient-occident premise. As that premise provides an opportunity for national identification through the metanarrative of multiculturalism, so the tension that Frye identifies constitutes a similar opportunity through a discourse about inter-regional dynamics in Canada. These two elements function together to provide a recognizably Canadian inflection on this conventional sitcom structure.

Little Mosque most often deploys the regionalist premise to interrogate attitudes pertaining to taste and cultural capital that reflect larger economic and cultural power imbalances between Canadian regions. In season one, episode 1, "Little Mosque," Amaar is astonished to learn that he cannot get a cappuccino at Fatima's cafe.[14] Annoyed by Amaar's supercilious attitude, Fatima pulls out a tub of ice cream and plops a scoop in Amaar's coffee: "There is your cappuccino!" In the same episode, Amaar makes an appearance on Fred's show in order to explain his presence in Mercy after his double foreignness – as a suspected Muslim terrorist and Torontonian – creates a stir. Fred goads him with slurs like "Johnny Jihad" and "Bedouin Buckaroo" until Amaar confesses his intense distaste for rural prairie life. He almost returns to Toronto, but Rayyan ultimately convinces him to stay.

As the series progresses and Amaar begins to settle into both Mercy and his role as an imam, he effectively becomes localized. In fact, when the Anglican parish receives a new reverend in season four, it is Amaar who emerges as a representative of the local community. Reverend Thorne comes from Toronto and re-ignites the regionalist aspect of the program's thematics with his references to high culture and his obvious contempt for Mercy. In 4.05, "Death by Chocolate," Thorne reads a weeks-old newspaper from Toronto. When this is pointed out to him, he exclaims, "Yes, but it's from Toronto and any news from Toronto is good news," while Fatima and the viewer see a large headline about an ongoing garbage strike. This bit of ironic commentary effectively encapsulates the relationship between Canada's centre and its peripheral Midwest; the marginal group is permitted to enjoy a laugh at the expense of the dominant entity and thereby reduces its stature.

In each of these cases, *Little Mosque* inflects a standard premise of the conventional sitcom in a manner that alludes to both established themes in Canadian nationalism and the CBC's core values. There is also a third inflection strategy that resonates with CBC traditions, which involves the incorporation of informational or educational content into a program.[15] *Little Mosque*'s persistent desire to inform its viewers about Islam is consistent with this practice. The program seems to presume that the viewer will be unfamiliar with the religion and its culture, and thus provides a great deal of information about it. This practice traces back to a long tradition in Canadian public broadcasting of producing informational programming via the National Film Board and CBC and a broader national interest in information-based programming (Hogarth 2002).[16] Even in fictional programming, there has been an interest in conveying information, addressing complex social issues, and portraying morals that are consistent with national ideals (Byers 2008). For example, Byers (2008, 190) argues that the *Degrassi* franchise emphasizes social and political issues in conjunction with a realist aesthetic in a manner that creates an "intertextual meta-language" for the franchise within the Canadian television industry and mediascape. This pedagogical approach to social issues and realism builds an identifiable and distinctive style that is articulated to the Canadian nation through its allusions to historically dominant discursive values in national public television. These attributes relate back to the aforementioned tradition of informational news, documentary, and realist drama programming.

Little Mosque labours in a similar manner through its deployment of pedagogical content in the sitcom context. The program reinforces established national values associated with multiculturalism and active citizenship through its emphasis on tolerance for others and the importance of doing the work to learn about other cultures and recognize and appreciate their attributes in a meaningful way. The series augments this broader thematic content with micro-lessons on Muslim culture that are woven into *Little Mosque*'s plotlines in a manner that privileges purposeful dialogue. The series' pilot 1.01 "Little Mosque" exemplifies this. Over the course of the episode, a traditional Muslim greeting and the appropriate response, the circumstances surrounding the initiation of Ramadan and its observance, and the manner in which Muslims pray are all incorporated into the plot. For example, in a key scene, several characters discuss whether Amaar should be permitted to use a telescope to sight the new moon that signals the beginning of

Ramadan, or whether he must use the naked eye. Similarly, in 1.05 "The Convert," the Muslim congregation is initially overjoyed at the prospect of its first Christian convert, but his ultra-orthodox ways soon wear on the moderate members of the community. In order to run him off, the Muslims perform an elaborate ruse involving all sorts of Western practices that are not permitted in Islam. The viewer learns that Muslims must not eat pork, drink alcohol, or gamble, and that Muslim men should not wear gold or silks. In season 3, during Rayyan's courtship with the engineer J.J., the viewer learns about the various practices and customs associated with dating and marriage in Islam. When considered as a whole, these instances amount to a significant educational component that distinguishes *Little Mosque* from most conventional American sitcoms.

With these different forms of inflection in mind, *Little Mosque*'s subtle variations on the sitcom form produce a text that might be read as Canadian when it is viewed as a CBC product in relation to other programming options in the Canadian telescape. Although it is a conventional sitcom in many ways, *Little Mosque*'s emphasis on the interplay between regions, multiculturalism, and informational content all situate it within established CBC television broadcasting practices. At the same time, these elements are sufficiently subtle that they would likely not be identified as overly localized – i.e., obtrusively Canadian – by viewers in other contexts.

The unobtrusiveness of these national qualities is undoubtedly useful to the program's producers as it facilitates the show's syndication in international markets even as its Canadian elements allow the producers to tap into lucrative funding opportunities at home.[17] At the same time, the subtle presentation of these national inflections is also a smart move when courting Canadian viewers. While the evidence suggests that the much-lamented distaste of Canadians for domestically produced content has been overstated, it is still the case that Canadian tastes in situation comedy have developed through exposure to foreign (and primarily American) content. The particular form of ambivalence that occurs when a program deploys national elements to inflect an externally defined genre creates a palatable national statement that achieves its full potency when aired in a national public broadcasting context.

SATIRE, PARODY, AND THE CULTURAL AMBIVALENCE OF *LITTLE MOSQUE*

The cultural ambivalence identified above also materializes in the program's deployment of parody and satire, albeit in a somewhat different form. These forms of comedic expression have a long history in Canada. Beverley Rasporich (2006) and Gerald Lynch (2007) trace this back to the humourist Stephen Leacock, if not his predecessors. Canadian television producers have picked up this literary tradition. Sketch comedy programs like *SCTV*, *CODCO*, *The Kids in the Hall*, *The Royal Canadian Air Farce*, and *This Hour Has 22 Minutes* solidified the pre-eminence of satire and parody as vehicles for Canadian national identification and cultural expression late in the mass-media era.

How might we account for this? Serra Tinic effectively encapsulates the appeal and utility of these strategies for Canadians in terms of the processes of negotiation involved in working out Canadian spaces in an Anglo-American context dominated by the US and Britain. She suggests, "As a form of cultural bonding, satire, and its relationship to power and marginality, is particularly effective as a mode of symbolic resistance to the perceived power of the dominant other" (2009b, 169). In terms of Canadian television, it is clear that irony and satire have been deployed in various ways to critique and subvert the dominant entities within the country on a regional level and a continental level in relation to the United States.

At the same time, parody has been a complementary discursive mode for Canadian cultural producers. In fact, parody often functions as intertextual shorthand that supports dominant or sanctioned readings by playing to the reader's textual or genre literacy (Gray 2006, 35–40). Katarzyna Rukszto (2005) and Zoë Druick (2008) each cite Linda Hutcheon's observation that parody can be both transgressive and supportive, depending on the orientation of the reference and the viewer's ability to recognize its intertextual dimensions. Hutcheon suggests that parodies legitimate their original texts to a certain degree because their transgressions are always authorized by the norms they seek to subvert. She contends that satire is concerned with moral, social, and political dimensions while parody is concerned with the aesthetic elements of another discursive text (cited in Druick 2008, 109). Satire thus reduces the stature of dominant entities, while parody often refers to shared cultural materials and frames of reference. Rukszto (2005, 77) writes, "In particular, parodic satire is well suited for exposing internal divisions, debunking dogma and making fun of institutional authority. This is why some suggest that parody coupled with satire is the most appropriate way to reflect the paradoxes of being Canadian."[18] Rukszto's implication is that parody without satire is effectively affirmative, whereas the satirical element brings with it the potential for critiques that might subvert dominant norms and thereby provide opportunities for distinction. In the case of the North American sitcom, parody is further complicated by virtue of its historical role as a key convention of North American television (Mellencamp 1992, 61). The use of parody in this context thus reaffirms *Little Mosque*'s consistency with its genre when specific and targeted forms of satire are not in the foreground. *Little Mosque* does not engage in a structural parody of the sitcom genre. Rather, its use of parody is consistent with the conventions of that genre.[19]

Little Mosque reflects the manner in which this dynamic often manifests in Canadian television programming. On the one hand, it deploys satirical characterizations and ironic humour to critique both Canadian regionalism and Western stereotypes about Islam that are subtly articulated to the United States. It is thus able to discursively undermine dominant entities on the national and continental levels. At the same time, it deploys genre and text-specific parodies in conjunction with references to American popular culture. Together, these elements reaffirm the program's grounding in a cultural context that is dominated by the United States. While *Little Mosque*'s parodic satire has the capacity to critique normative values and provide opportunities for national distinction, its less critical parodic references merely function intertextually to drive the comedic action. They thus reaffirm the importance to Canadians of a shared repository of mass media texts that has largely been produced by the American cultural industries. In other words, if the satirical elements provide opportunities for critical distinction along national lines, the straightforward parodies reassert the importance of a shared Anglo-American popular cultural heritage that the United States has overdetermined.

While Druick observes that sketch comedy programs on Canadian television have frequently operated in this manner, *Little Mosque* engages in these practices in the situation comedy context. The program devotes sustained satirical consideration to the predominant

Western representations of Muslim fundamentalism, of the media institutions that propagate these stereotypes and create suspicion, and the citizenry that internalizes these messages and then acts on them. In this context, these representations are implicitly articulated to the United States, the dominant force within the Western bloc that has so long privileged these Orientalist conceptions of the East and the East-West dichotomy they sustain. In conjunction with the content discussed in the previous section, these depictions provide a point of distinction for Canadian viewers that might encourage national identification on the basis of Canada's multicultural ideals.

The term *stereotype* has been defined as "an exaggerated belief associated with a category ... its function is to justify conduct in relation to that category" (Gordon Allport, cited in Noriega 2000, 30). Chon Noriega notes that the crucial issue here is the "social function" of the stereotype. In the West, the stereotypes outlined in Edward Said's (1979) *Orientalism* still retain a significant amount of utility for many people. The Orientalist construct gives the imperialist West a certain power over the East, but it also constructs the Orient as both threatening and exotic in various guises. These images are still pervasive in the West. Muslims are often considered to be an unassimilable other: they are presented as too different to blend into the American melting pot when they are not caricatured as violent extremists who wish to harm America. Research indicates that the latter representation has long predominated in the depictions of Muslims produced by America's cultural industries (Shaheen 2001; 2008). In fact, Karim (2000, 1) posits that Islam has become the West's "primary other" in the aftermath of the end of the Cold War. Given the nature of this event, and the United States' hegemonic status within the global West, it figures that the United States is often primarily associated with this discursive framework despite the documented existence of significant amounts of prejudice in places like Canada and the United Kingdom. Although critics have argued that satirical commentary on US politics became less palatable in Canada after the events of 9/11 (Rasporich 2006, 58; Tinic 2009b, 177–8), *Little Mosque* suggests that either more subtle forms of satirical critique continue to be palatable or that the climate is shifting back towards more aggressive forms of satire as that cataclysmic event recedes into the past.

Little Mosque invites viewers to consider their relationships to these stereotypes and their attitudes towards Islam, but also permits them to enjoy the (somewhat fanciful) notion that such stereotypes belong to the United States and not to the ostensibly multicultural Canadian nation – this is the aforementioned fantasy of difference. These depictions present Canadian viewers with opportunities for negative self-identification. Canadian viewers may feel superior by virtue of their tolerant, liberal-pluralist values as they watch the program and associate stereotypes with the United States.

The program primarily critiques these Orientalist stereotypes through three characters: the conservative Muslim elder Baber, the irascible and prejudiced radio personality Fred Tupper, and the paranoid and ignorant farmer Joe. If Baber is a satirical parody of the stereotypical image of the Muslim fundamentalist, then Fred represents the media outlets that propagate such images, and Joe stands for those who internalize them and turn them into fear and exotic fascination.[20]

Baber is a conservative economics professor and single father who believes that the "beardless imam" Amaar is too liberal to lead the congregation (Baber thinks he would be a more appropriate alternative). Baber subverts the stereotype of the West-hating Muslim extremist through his curmudgeonly and overly fastidious approach to life. At the same time, Baber also holds extremely conservative views about gender and sexuality; on that level, he does not so much subvert stereotypes as represent more orthodox views within the Muslim community. Baber's diatribes frequently concern personal gripes or inconsequential issues. For example, he frequently rants about Western decadence, but these selective tirades often focus on American popular culture rather than on the core values of liberal democratic societies. In fact, Baber sometimes ventriloquizes long-standing Canadian cultural nationalist arguments against debasing American mass culture in a manner that falls somewhere between endorsing and lampooning them. For example, in 1.01, Baber delivers a sermon in which he excoriates the Idol franchise and *Desperate Housewives*. He shouts, "*American Idol! Canadian Idol!* I say, all idols must be smashed!" This is a clever play on both the common conception that *American Idol* is a crassly commercialized, lowbrow mass product and the sense that the Canadian incarnation of the franchise does not reflect the nation in any meaningful way, which alludes to long-standing concerns that Canada is merely a branch plant for the American cultural industries. He proceeds to address the ABC hit *Desperate Housewives*, wondering, "Why are they so desperate? They are only fulfilling their womanly duties." Meanwhile, in the congregation, Rayyan and Sarah Hamoudi whisper about the most recent episode. This interchange is an ideal example of the collective Canadian ambivalence towards American popular culture under the guise of critiquing conservative Muslim attitudes towards the role of women in the home and society.

Baber's case is complicated, however, by his devotion to *Coronation Street*, a long-running British soap that airs weekday afternoons on the CBC, a fact of which Canadian viewers would likely be aware. By presenting the conservative Muslim extremist as a grumpy old economics professor with a fusty disdain for contemporary American popular culture, *Little Mosque* subverts the West-hating Muslim extremist stereotype. At the same time, Baber's Anglophilic tendencies underscore the manner in which his anti-Western rants resonate with long-standing Canadian nationalist discourses. By satirically configuring Baber's extremism in a way that suggests Anglophilic disdain for lowbrow, mass culture that is coded as American, the program lets Canadian viewers in on the joke, which absolves them of blame for the stereotype and gives them an opportunity to dissociate themselves from it.

Baber's natural counterpart within the Christian community is Fred Tupper, the host of the local radio talk show *Wake Up, People!* Fred frequently caricatures and criticizes the Muslim community on his program, thereby creating suspicion and encouraging prejudice within the community. His characterizations register as hyperbolic and outlandish to viewers who likely see him as a satirical commentary on the hate-filled invective spewed by conservative talk show hosts in North America. In 1.01, Fred interviews Amaar on his program and bombards him with inflammatory accusations and discriminatory characterizations before telling him, "Please, feel free to give as good as you get. That's the privilege of living in a country with freedom." This conception of freedom as the ability to return verbal fire likely resonates with most Canadian viewers as associated with the United States. The emphasis on individual freedoms – particularly freedom of speech – in this exchange speaks to fundamental American values. This underscores the manner in which Fred provides Canadian viewers with an opportunity to distinguish themselves from the United States via a satirical commentary on those who invoke militant or confrontational conceptions of freedom in order to justify suspicion and discrimination. Although Canada has its share of these personalities, Canadians might view this character as American given the association of this sort of programming with the American right.[21]

This is only half the story with Fred, however. Although his radio persona emphasizes belligerent prejudice towards the Muslim community, he develops an increasingly close relationship to it in his off-air life. He frequently eats at Fatima's cafe and feelings appear to develop between the two of them. Rayyan serves as his doctor and he often socializes with members of the Muslim community. This situation initially seems incongruous, though Fred admits that he demonizes the Muslim population primarily because it is good for his program's ratings. In 2.12 "Jihad on Ice," Fred says this outright. In 2.17 "Meet J.J.," he endeavours to avoid this practice out of respect for his new intern Layla, only to see his ratings take a precipitous drop.[22] The dynamic is perhaps best illustrated by 3.20 "Can I Get a Witness?" wherein Fred insists on attending Rayyan and J.J.'s wedding. There, he delivers a heartfelt impromptu speech at the reception after J.J.'s departure before insisting that he must go off to prepare the next day's radio program. When Amaar inquires as to the program's theme, Fred responds, "Crazy Muslim weddings," revealing that he intends to offer a misleading portrait of the event in order to pander to his non-Muslim audience.

Fred represents ignorance, hypocrisy, and crass commercialism – values that have long been associated with the United States in Canadian cultural nationalist discourse. He is initially portrayed as suspicious and prejudiced as a result of his lack of familiarity with Islam and the local Muslim community. As he becomes increasingly involved with that community, viewers learn that his ongoing tirades on his radio program are a function of commercial interest and not earnest sentiment. He thus resonates as a satirical characterization of the right-wing media in the United States, which allegedly plays to its listeners' fears and prejudices as a means of gaining market share. As Sandra Cañas (2008, 199) observes, "The radio pundit shows how the media operates as a site of production of orientalist imaginaries, which are later disseminated throughout and internalized by groups in society, such as the disturbed white man who calls the 'terrorist hotline' after having come across people praying." This is an accurate characterization, but this image of "the media" has been overdetermined by the United States and may not necessarily carry these connotations in the Canadian context.

This "disturbed white man" is the third parodic figure in the form of Joe, the unsophisticated farmer who lacks worldliness and common sense. Like Fred, Joe provides a potential mechanism for negative self-identification because he is presented as an ignorant, suspicious, and unintelligent caricature. Joe's conservative values also deviate from those that are often considered normative in Canadian society (though the actual norm is probably somewhat closer to Joe's position). He stands as a point of potential distinction on both regional and national levels, shoring up ideals about the Canadian nation through his embodiment of rural lack of sophistication and his espousal of ignorant views that often ventriloquize suspicions of Muslims and Islam that Canadians often associate with the United States.

The program does not typically invite viewers to identify with Joe (though in later seasons he reveals a kind, sweet side). Instead, he functions as a means to show how ridiculous and pernicious these stereotypes can be when they are acted on. Joe represents those individuals who have little contact with Muslims, yet develop strong beliefs about them through media coverage. In 1.01, he telephones a terrorist hotline after stumbling upon the Mercy Muslims in the midst of a prayer session. Discussing the matter with Magee, Joe says, "I saw them bowing and moaning just like on CNN," which illustrates the role of the mainstream American news media in the propagation of negative images of Muslims.

Joe and Baber often serve as comedic and thematic foils for each other. In 2.02 "Ban the Burqa," a burqa-clad woman begins to worship at the mosque. Baber becomes romantically interested in her because of her piety while, in a perfect depiction of the allure of the East within the Orientalist imaginary, Joe develops an exotic fascination with her. Joe ultimately dons a burqa in order to pursue and expose the mysterious figure, while Baber pines for her from afar (until she introduces herself and he dismisses her for being immodest). Joe and Baber also form a strategic alliance over the issue of gay marriage in 1.07 "Mother-in-Law." They discover that they are the only individuals in the community who strongly oppose the idea and so together they organize a demonstration in protest of a planned marriage. When the rally fails to attract any additional citizens, the two head off together and Joe offers to buy Baber a beer. Of course, Joe does not appreciate that Muslims are not permitted to drink alcohol, but the moment nicely underscores the manner in which the shared conservative values of these two characters facilitate bonding. This amounts to an incisive ironic commentary on the dynamic between conservative Muslims and conservative Christians in the West.

These satirical parodies are always grounded in the Canadian context, despite the manner in which each one resonates with American attitudes and values. The point is not that these characters are always implicitly coded as American, but that by digesting them, Canadian viewers may disassociate themselves from those who believe some of the more negative conceptions of Islam in the West. Given that the US is the dominant force within the West and the primary producer of these images in a post-9/11 world, this provides an opportunity for distinction for Canadian viewers. In this case, Canada's marginal economic and cultural position within its globally dominant bloc potentially allows its citizens to deny their complicity with some of the more negative practices associated with that bloc. Canadians are allowed to embrace an ideal of sophistication and tolerance as the negative qualities are subtly articulated to the United States (and, in the case of Joe, rural Canadians).

If *Little Mosque*'s deployment of satirical critiques helps to establish Canada's ex-centricity in relation to these negative Western attitudes and stereotypes that are coded as American, its uses of parody involve more direct instances of intertextuality that draw on established genres and canonical texts in North American film and television. If the satirical elements provide opportunities for distinction, the parodic allusions re-assert the importance of a US-dominated repository of shared popular texts. Considering these parodic allusions in relation to *Little Mosque*'s extensive references to American programs underscores the permeability of the border between the Anglo-Canadian and American mediascapes for the Canadian viewer.

The classic American boxing film *Rocky* is a consistent parodic resource for *Little Mosque*. In 2.16 "Jihad on Ice," Baber reacts to Amaar's desire to form a curling team by obtaining a book on the sport and learning everything about it. He reads an exceedingly large volume on the subject as he walks around Mercy; the background music signals that the scene is a parody of the famous training montage in *Rocky* where the protagonist runs around the city and finishes on the steps of the Philadelphia Museum of Art. In this case, Baber finishes the book on the steps of a municipal building and raises it above his head in triumph. This is no critique, but merely a humorous intertextual moment that ridicules Baber's method of learning about curling while it indirectly foreshadows the key role that his arcane knowledge will play in the outcome of the big bonspiel. A reasonable facsimile of *Rocky*'s well-known showdown song "The Eye of the Tiger" plays just prior to the pivotal quiz confrontation between Baber and Reverend McGee in 3.15 "Colour Me Excited." Again, this is an ironically humorous allusion that establishes the scale of this "fight" and disproportionate level of the combatants' investment in it. Finally, in 4.09 "Gloves Will Keep Us Together," there is a parody that verges on homage when "The Eye of the Tiger"-like music plays once again as Reverend Thorne (the "Minister of Justice") and Amaar take to the ring for a charity boxing match. The aural allusion here once again foreshadows the underdog's victory as Amaar pulls out an unexpected TKO. In each of these examples, the program depends on the audience's presumed familiarity with the *Rocky* franchise for comedic effect. This is not the biting irony that allows the marginal entity to cut the dominant down to size, but a lighter variant that uses an affirmative form of intertextuality to drive the comedic action by highlighting the characters' pretences.

There are other such parodies as in 1.05 "The Convert," where Baber dresses up as "Mr B," a Mr T parody intended to help drive away the over-zealous convert Marlon, and 4.12 "Pants on Fire," where Mayor Popowicz tells her assistant Sarah, "You complete me," in an allusion to a pivotal scene in the Tom Cruise film *Jerry Maguire*, as they reconcile after a brief falling out. In each of these instances, the action depends on the viewers' familiarity with key texts in recent American popular culture. Rather than subverting them, it merely uses them for comedic effect.

Little Mosque also trades regularly in genre parodies. For example, episode 2.02 "Ban the Burqa" features an ironic parody of a film noir confrontation between Magee and Amaar pertaining to the latter's inability to effectively hide a box of fudge. Episode 3.11 "True Bromance" plays on the buddy film genre in a similar fashion as Nate, the reporter at the town newspaper, befriends Amaar and convinces him to stop at a local roadhouse for some hot wings, which leads to predictable comedic misadventures. These genre parodies draw on a mass media heritage that Canada shares with the United States, although the United States dominates and defines it. In line with the conventions of the sitcom, *Little Mosque* parodies these genres in order to quickly convey information to its viewers and produce comedic effects. Where more satirical elements that pertain to questions of cultural representation and power produce opportunities for distinction, these opportunities subtly reaffirm that Canadians depend on a US-dominated Anglo-American popular culture heritage. This point is perhaps made most clearly in a brief scene from 3.16 "Recipe For Disaster," in which Fred and Layla find Fatima's cafe closed and vow to work together to find out what has happened. The scene concludes with the following exchange, in which Fred endeavours to make sense of their parodic partnership in terms of investigative duos from bygone eras in network television:

Layla: Don't worry; we'll get to the bottom of this.
Fred: Yeah, yeah ... we'll be like Starsky and Hutch.
Layla: Who's that?
Fred: Hardcastle and McCormick.
Layla: (*Frowns in incomprehension.*)
Fred: Uh ... McMillan and wife.
Layla: Oh, I love that show! (*Departs.*)
Fred: (*Smiles to himself.*) Aren't they great?

Here two characters, the Muslim-disparaging conservative Christian Fred and the Muslim teenager Layla, form a brief, parodic partnership in order to find out what has happened to Fatima's cafe. The viewer sees this as an image of multicultural harmony on the individual and local levels. At the same time, the bonding exercise centres on Fred's attempt to make a comparison between himself and Layla and various famous detective partnerships from television's past. The momentary disjuncture here is not a function of cultural background, but of differing generational perspectives (though reruns of American programming have clearly helped level the field).[23]

While this is not a reflexive event, it underscores the manner in which the program draws on US popular culture to drive its humour and dramatic action. The gag initially hinges on Layla's lack of awareness of two programs that pre-date her birth by one or more decades. It is only with the relatively obscure Rock Hudson–Susan Saint James vehicle *McMillan & Wife* (NBC, 1971–77) – a somewhat desperate third attempt by Fred – that Layla recognizes a reference. This, in turn, registers as entertaining or interesting to viewers by virtue of their familiarity with these references, which in this case would require knowledge of American network television detective series from the 1970s and 1980s. In other words, the gag at the heart of the scene uses a shared repository of US popular cultural texts with these references. Canadian viewers might be expected to know them and to derive pleasure from the comedic action they facilitate.

These examples, and this broader discussion, speak to an ambivalence that exemplifies the Canadian mass media experience. The satirical critiques of Muslim stereotypes are certainly not confined to the United States – after all, they are set in Canada, which represents the generic West – but the normative values and nationalist metanarratives at play in both countries might encourage Canadian audiences to associate them with the United States. If these elements facilitate distinction along national lines, the text and genre parodies and cultural references draw Canada back into a US-dominated Anglo-American cultural framework. *Little Mosque* thus writes across the border – it agonistically negotiates between elements coded as American and those coded as Canadian as its story unfolds from episode to episode.

CONCLUSION

The tension between these satirical and parodic elements in *Little Mosque* in conjunction with the program's localized inflections of externally defined genres, themes, and situations reflects Anglo-Canada's position between Great Britain and the United States in Anglo-American culture. A marginal entity in a globally dominant bloc, the Canadian cultural experience is defined by a cultural ambivalence that results from the prevalence of and persistent need to engage with foreign elements in the Canadian mediascape. The result is the practice of localizing these elements through creative inflection and the tendency to establish national distinctiveness and legitimacy through the use of satire, which is countered by the affirmative intertextuality of gentle parodies of the sort found on *Little Mosque*. Instances of national distinction are useful here, but this process also involves the tacit acknowledgement that "American" elements are central to Canada's mass media heritage.

Levine (2009, 529) argues: "Canadian identity is most typically conceived in terms of vagueness or negation, as marginal or not-American, the very lack of specificity inherent in Canadianness allows the difference and distinctiveness of Canadian fare ... to be taken up and read in any number of ways." I submit that these variegated readings reflect the creative processes of cultural negotiation that appear in texts like *Little Mosque*. It may be true that "the political and cultural insecurity that accompanies the seemingly insubstantial border between the two countries provides perhaps the greatest sense of marginalization for most Anglo-Canadians" (Tinic 2009b, 170). However, as the preceding discussion of the relationship between marginality and the assertion of superiority illustrates, this feeling can be productive when it motivates Canadian producers, cultural intermediaries, and audiences to creatively engage with cultural elements that originated elsewhere. It is not vagueness about the nation in terms of its symbols, myths, and metanarratives, but the persistent need to engage with dominant genres, images, and narratives produced elsewhere that defines much Canadian cultural production.[24] As *Little Mosque* demonstrates, when Canadian themes and practices are deployed in order to inflect or subvert genres, representations, and themes associated with other national contexts, localized statements emerge through the contrast with elements marked as American even as those statements attest to the pivotal role of American popular culture in English Canada. *Little Mosque* also attests to the importance of the network context to this dynamic; the CBC's iconic status undoubtedly helps it produce cultural statements that can be read as Canadian. In the end, *Little Mosque* suggests that Canadian television does not so much involve "writing on the border" as negotiating between the regional, national, and international elements at play in the Canadian telescape (Berland 2001). If this practice suggests a certain cultural ambivalence, it also alludes to possibilities for identification through the sharp contrast between the marginal local and dominant foreign as they rub against one another in the same program and network contexts in a crowded Anglo-Canadian telescape.

NOTES

1 I am excluding Quebec and the rest of French-speaking Canada from this discussion despite the fact that this program has aired on SRC, the CBC's French-language broadcaster. While I acknowledge the need for more work that considers the Canadian experience in terms of the whole of Canada (English, French, First Nations, and other groups), it is beyond the capacity of this paper to address more than the English-speaking Canada.

My interest in the manner in which the CBC frames the program in the Anglo-Canadian telescape also partly motivates this exclusion.

2 Television is considered "Canadian" based on the nationalities of the producer and creative personnel, the extent to which Canadians were compensated for services related to the production of the programming, and whether or not paid lab processing occurred in Canada. The CBC must air at least 60 per cent Canadian content based on whether or not its producer and creators are Canadian, whether Canadians made the program, and the amount of money spent on processing in Canada. At least 60 per cent of the CBC's broadcasting schedule must be Canadian between 6 a.m. and 12 a.m. Private broadcasters must maintain the 60 per cent rate over the course of the day, but may dip down to 50 per cent from 6 p.m. to 12 a.m.

3 As Bodroghkozy points out, Miller's essay is itself part of a collection that specifically addresses the question of how Canadian producers and users engage with American popular culture.

4 To date, the program has been syndicated in a variety of foreign countries, including Finland, France, Turkey, the United Arab Emirates, and a number of North African countries. Fox purchased the American rights, though it has yet to air the program or produce the planned American version. In Canada, the program debuted with an incredible 2.1 million viewers, though weekly ratings slid to around 1 million viewers by the end of season one and had slipped to 400,000 viewers by season four.

5 In a more recent piece, Liz Czach (2010) identifies a productive turn towards content analysis in Canadian television studies, though she also maintains that more work needs to be done in this area.

6 Nawaz is perhaps best known for her feature-length documentary *Me and the Mosque* (2005), which draws on Nawaz's own experiences to examine the place of women in Islam in the contemporary Canadian context with an historical perspective.

7 Canadian content regulations and US dominance in many entertainment genres also contributed to this situation as news programming is fundamentally local and cannot easily be imported.

8 Extant genre hierarchies in television may also contribute to this. Paul Attalah (2002) contends that the sitcom has often been considered the "unworthy discourse" in relation to more valued genres like drama and documentary.

9 Another recent essay on the program suggests that *King* is far more critical than it has generally been considered to be. Jennifer Vanderburgh (2008)

contends that the program goes beyond merely affirming the multicultural ideal to interrogate the issues beneath that ideal.

10 Elana Levine (2009) also notes the presence of this dichotomy in her discussion of the *Degrassi* franchise's emphasis on the liberal values of multiculturalism and tolerance. Levine's essay is particularly valuable because of the manner in which she attempts to discursively acclaim these programs as Canadian on this basis.

11 See Day (2000) and Mackey (2002). Mackey argues that Canada's multiculturalism emanates from a British benevolence to "others" and that it works to structure difference in Canadian society, preserving existing social hierarchies and precluding the meaningful recognition of difference. Day similarly argues that Canada's multicultural metanarrative has obscured tendencies toward assimilation throughout Canada's history.

12 The Broadcasting Act, 1991 states, among other requirements, that the programming provided by the CBC should "reflect the multicultural and multiracial nature of Canada." The act also emphasizes the CBC's role in producing a national consciousness and facilitating a dialogue among Canada's regions.

13 Bodroghkozy (2002, 576) goes so far as to contend that Toronto sometimes stands as "America north of the 49th parallel" when this centre-periphery dynamic is invoked on the national level.

14 From this point forward, I will use the abbreviation X.X to indicate the season and episode number, e.g., 1.01 for the series' pilot.

15 Matheson (2012) calls this practice "demystification." Her analysis centres on 1.03, "The Open House," in which the mosque opens its doors to the broader community and the Muslim characters teach their visitors – and the viewers at home – about Islam.

16 Aniko Bodroghkozy (2002, 575) relates that the news anchor Knowlton Nash once referred to Canadians as "infomaniacs" for their perceived collective investment in information-based programming.

17 International syndication has been highly lucrative for Canadian producers in recent years. Tinic (2005) and Levine (2009) both conduct detailed discussions of this phenomenon.

18 It is perhaps becoming more difficult to separate satire and parody. Druick contends that satire invariably involves a commentary on the world, whereas parody involves a commentary on texts. Yet, in our increasingly "hyperreal" world, this distinction may not be as useful as it was in previous eras.

19 In contrast, Jonathan Gray's (2006) analysis of *The Simpsons* suggests that the program actively deploys parody to critique the sitcom genre.

20 It is important to note here that a number of television critics and media observers have argued that Joe's character constitutes a one-dimensional portrayal of rural Canadians as ignorant and prejudiced individuals, and thereby perpetuates a pernicious urban/rural dichotomy in Canada.

21 Influential Canadian television critic John Doyle affirms this, and states "the kind of redneck attitudes from some of the locals in the small town toward the Muslims is very much reflective of an American suspicion of Muslims and not a Canadian suspicion" (quoted in Richard Reynolds, NPR, 18 January 2007). Similarly, Mark Steyn's controversial assessment of the program involves an assertion that, in real life, Tupper would be hauled before the Saskatchewan Human Rights Commission. The implication is that his act is not consistent with the accepted practices and values in Canadian society (*Maclean's*, 5 February 2007). Finally, Margaret Wente opines that the Canadian Radio-television and Telecommunications Commission would have removed Tupper from the air for the views he espouses, though she also notes that the program sets him up as a sympathetic character on a personal level (*Globe and Mail*, 9 January 2007). The critical point here is that this character is not understood to correspond to the normative values held by Canadians and Canadian media observers and that he is often understood implicitly or explicitly in terms of the United States.

22 In fact, Joe accuses Fred of "going soft" when Fred decides to stop criticizing the Muslim community on his program.

23 This moment stands in pointed contrast to Canada's own televisual history. While Canadian viewers have ample access to American programming reruns, domestic programming from past eras is much more difficult to find. This has prompted Serra Tinic (2009a) to describe Canada as a "no-rerun nation."

24 I write this in reference to the hegemonic forms of Canadian cultural nationalism that tend to predominate on the CBC. My intention is not to endorse them, but to note that for many Canadians these values form the basis for a coherent and stable national subjectivity that departs significantly from the "crisis" and "uncertainty" tropes that seem to consistently recur in writing about Canadian identity. Furthermore, when I allude to how feelings of marginality can be "productive," I do not intend to suggest that this is invariably a positive phenomenon, but merely that it can result in coherent nationalist statements that are of utility to some in the population.

REFERENCES

Alsultany, Evelyn. 2008. "The Primetime Plight of the Arab Muslim American after 9/11." In *Race and Arab Americans Before and After 9/11*, edited by Amaney Jamal and Nadine Naber, 204–28. Syracuse, NY: Syracuse University Press.

Anderson, Benedict. 1991. *Imagined Communities: Reflections on the Origins and Spread of Nationalism*. London: Verso.

Attalah, Paul. 2002. "The Unworthy Discourse: Situation Comedy in Television." In *Critiquing the Sitcom: A Reader*, edited by Joanne Morreal, 91–115. Syracuse, NY: Syracuse University Press.

Beaty, Bart, and Rebecca Sullivan. 2006. *Canadian Television Today*. Calgary: University of Calgary Press.

Berland, Jody. 1996. "Space at the Margins: Colonial Spatiality and Critical Theory after Innis." *Topia* 1: 55–82.

– 2001. "Writing on the Border." *CR: The Centennial Review* 1 (2): 139–69.

Bodroghkozy, Aniko. 2002. "As Canadian as Possible …: Anglo-Canadian Popular Culture and the American Other." In *Hop on Pop: The Politics and Pleasures of Popular Culture*, edited by Henry Jenkins, Tara McPherson, and Jane Shattuc, 566–88. Durham, NC: Duke University Press.

Byers, Michele. 2007. "*Little Mosque on the Prairie*: The Life and Times of the CBC." *Flow*, 23 February. http://www.flowtv.org/2007/02/little-mosque-on-the-prairie-the-life-and-times-of-the-cbc.

– 2008. "Education and Entertainment: The Many Reals of *Degrassi*." In *Programming Reality: Perspectives on English-Canadian Television*, edited by Zoë Druick and Aspa Kotsopolous, 187–204. Waterloo: Wilfred Laurier University Press.

– 2009. "Our Little Mosque." *In Media Res*, 6 April. http://mediacommons.futureofthebook.org/imr/2009/04/05/our-little-mosque.

Cañas, Sandra. 2008. "The Little Mosque on the Prairie: Examining (Multi)Cultural Spaces of Nation and Religion." *Cultural Dynamics* 20: 195–211.

Charland, Maurice. 2004. "Technological Nationalism." In *Communication History in Canada*, edited by Daniel J. Robinson, 28–39. Don Mills, ON: Oxford University Press.

Czach, Liz. 2010. "The 'Turn' in Canadian Television Studies." *Journal of Canadian Studies* 44 (3): 174–80.

Day, Richard F. 2000. *Multiculturalism and the History of Canadian Diversity*. Toronto: University of Toronto Press.

Druick, Zoë. 2008. "Laughing at Authority or Authorized Laughter? Canadian News Parodies." In *Programming Reality: Perspectives on English-Canadian Television*, edited by Zoë Druick and Aspa Kotsopoulos, 107–28. Waterloo: Wilfrid Laurier University Press.

Druick, Zoë, and Aspa Kotsopoulos. 2008. "Introduction." In *Programming Reality: Perspectives on English-Canadian Television*, edited by Zoë Druick and Aspa Kotsopoulos, 1–14. Waterloo: Wilfrid Laurier University Press.

Edwardson, Ryan. 2008. *Canadian Content: Culture and the Quest for Nationhood*. Toronto: University of Toronto Press.

Frye, Northrop. 1971. *The Bush Garden: Essays on the Canadian Imagination*. Toronto: Anansi.

Gray, Jonathan. 2006. *Watching with The Simpsons: Television, Parody, and Intertextuality*. New York: Routledge.

Hogarth, David. 2002. *Documentary Television in Canada: National Public Service to Global Marketplace*. Montreal: McGill-Queen's University Press.

Karim, Karim H. 2000. *Islamic Peril: Media and Global Violence*. Montreal: Black Rose.

Khan, Sarah. 2009. "The Many Faces of Muslim Women in Canada: A Re-constructed Image in CBC's *Little Mosque on the Prairie*." Thesis, University of Ottawa.

Levine, Elana. 2009. "National Television, Global Market: Canada's *Degrassi: The Next Generation*." *Media, Culture and Society* 31: 515–31.

Lynch, Gerald. 2007. "Canadian Comedy." In *Comedy: A Geographic and Historical Guide*, edited by Maurice Charney, 199–213. New York: Praeger.

Mackey, Eva. 2002. *The House of Difference: Cultural Politics and National Identity in Canada*. Toronto: University of Toronto Press.

Manning, Frank. 1993. "Reversible Resistance: Canadian Popular Culture and the American Other." In *The Beaver Bites Back? American Popular Culture in Canada*, edited by David H. Flaherty and Frank Manning, 2–28. Montreal and Kingston: McGill-Queen's University Press.

Matheson, Sarah A. 2006. "Ruling the Inner City: Television, Citizenship, and the *King of Kensington*." *Canadian Journal of Film Studies* 15 (1): 47–62.

– 2012. "Television, Nation, and Situation Comedy in Canada: Cultural Diversity and *Little Mosque on the Prairie*." In *Canadian Television: Text and Context*, edited by Marian Bredin, Scott

Henderson, and Sarah A. Matheson, 153–72. Waterloo: Wilfrid Laurier University Press.

Mellencamp, Patricia. 1992. *High Anxiety: Catastrophe, Scandal, and Comedy*. Bloomington: Indiana University Press.

Miller, Mary Jane. 1993. "Inflecting the Formula: The First Seasons of *Street Legal* and LA *Law*." In *The Beaver Bites Back? American Popular Culture in Canada*, edited by David H. Flaherty and Frank Manning, 104–22. Montreal and Kingston: McGill-Queen's University Press.

Noriega, Chon A. 2000. *Shot in America: Television, the State, and the Rise of Chicano Culture*. Minneapolis: University of Minnesota Press.

Pevere, Geoff, and Greig Dymond. 1996. *Mondo Canuck: A Canadian Pop Culture Odyssey*. Toronto: Prentice-Hall.

Rasporich, Beverly. 2006. "Canadian Humour and National Culture: Move Over, Mr Leacock." In *Canadian Cultural Poeisis: Essays on Canadian Culture*, edited by Garry Sherbert et al., 51–66. Waterloo: Wilfrid Laurier University Press.

Rukszto, Katarzyna. 2005. "The Other Heritage Minutes: Satirical Reactions to Canadian Nationalism." *Topia: A Canadian Journal of Cultural Studies* 14: 73–91.

Said, Edward. 1979. *Orientalism*. New York: Random House.

Shaheen, Jack. 2001. *Reel Bad Arabs: How Hollywood Vilifies a People*. Brooklyn: Olive Branch Press.

– 2008. *Guilty: Hollywood's Verdict on Arabs after 9/11*. Brooklyn: Olive Branch Press.

Stratton, Jon, and Ien Ang. 1998. "Multicultural Imagined Communities: Cultural Difference and National Identity in the USA and Australia." In *Multicultural States: Rethinking Difference and Identity*, edited by D. Bennet, 135–62. London and New York: Routledge.

Tinic, Serra. 2005. *On Location: Canada's Television Industry in a Global Market*. Toronto: University of Toronto Press.

– 2009a. "No Rerun Nation: Canadian Television and Cultural Amnesia." *Flow*, 12 June. http://www.flowtv.org/2009/06/no-rerun-nation-canadian-television-and-cultural-amnesia-serra-tinic-university-of-alberta.

– 2009b. "Speaking 'Truth' to Power? Television Satire, *Rick Mercer Report*, and the Politics of Place and Space." In *Satire TV: Politics and Comedy in the Post-Network Era*, edited by Jonathan Gray, Jeffrey Jones, and Ethan Thompson, 167–86. New York: New York University Press.

Urquhart, Peter, and Ira Wagman. 2006. "Considering Canadian Television: Intersections, Missed Directions, and Prospects for Textual Expansion." *Canadian Journal of Film Studies* 15 (1): 3–7.

Vanderburgh, Jennifer. 2008. "Imagining National Citizens in Televised Toronto." In *Programming Reality: Perspectives on English-Canadian Television*, edited by Zoë Druick and Aspa Kotsopoulos, 269–90. Waterloo: Wilfrid Laurier University Press.

3

The Flow of Amusement:
The First Year of Moving Pictures
in the Red River Valley

PAUL MOORE

Moving pictures debuted in the Red River Valley in early July of 1896 for an audience in Pembina, North Dakota, a small town at the international boundary of the forty-ninth parallel. As Thomas Edison's latest marvel and as a novelty of electric invention and a metropolitan entertainment, the location seems incongruous. Why not a city? Why did cinema arrive in Pembina before Fargo, Grand Forks, or Winnipeg? Pembina's curious historic significance can be seen as an extension of its legacy as a fur-trading post, one of the earliest European outposts in the region. The town was one of Lord Selkirk's settlements for the Hudson's Bay Company before the Treaty of Ghent established the international boundary at the forty-ninth parallel in 1818 (see Bumsted 2008). Pembina was still assumed to be within British North America until a boundary survey in 1823 revealed the town was in the United States, at which point the Hudson's Bay outpost moved to Fort Garry, near present-day Winnipeg. It seems fitting, then, that this border town happened to be the best site to unveil the modern amusement of cinema in the region, a marker of the international boundary's contradictory role in mediating transnational cultural connections. The first movie showmen in the Red River Valley had many ways to exploit the new amusement. Some collaborated with each other across the border, or crossed the border themselves, while others used moving pictures for nationalist purposes and considered the border a protective barrier. Neither political nor market boundaries mark this region entirely;

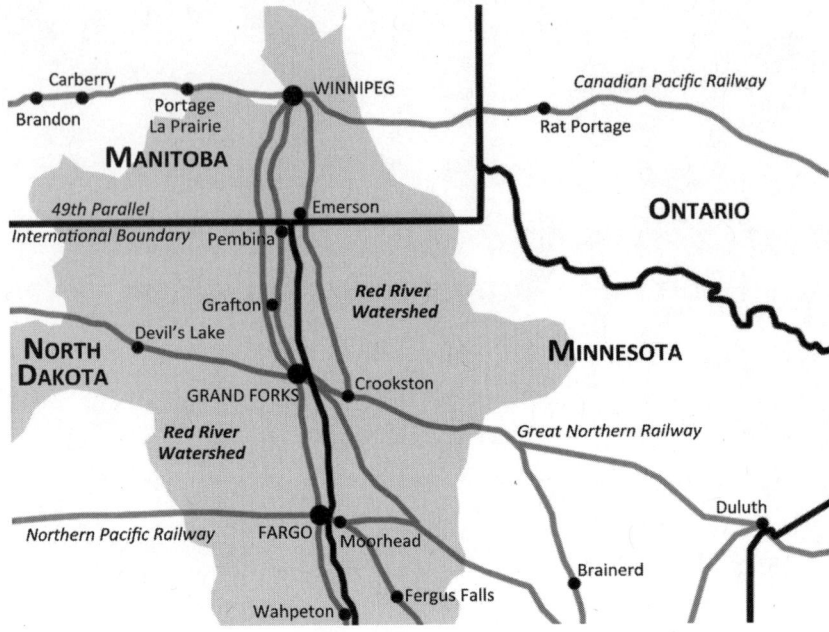

Figure 3.1 The Red River Valley (author)

instead, both influences meet within it. Astride political boundaries, the Red River Valley kept its residual legacy of north–south communication routes built along the river's path. Cinema showmen continued this legacy. Though at times they recognized the international boundary and at other times they ignored it, they added a regional north-south inflection to the dominance of the east-west traffic that enforced national differences.

The American national project can avoid or marginalize its northern frontier, but the forty-ninth parallel is an unavoidable determinant in histories of the Canadian Prairies (Stuart and Taylor 2005; Thompson and Randall 2008). Within the United States, studies across the international boundary that examine areas such as the Red River Valley in North Dakota and Minnesota have only regional importance, but within Canada, precisely the same studies are flooded with national significance when they examine Manitoba. The more codified clarity of Canadian national policy on the late-nineteenth century Prairies becomes a crucial starting point, then, in understanding how Winnipeg's significance as the east-west hub of the Canadian Pacific Railway superseded its location downstream on the north-south path of the Red River. Gerald Friesen (1987) centres the history of the Canadian Prairies on such policy cornerstones as treaties with aboriginal peoples, establishment of the Northwest Mounted Police, construction of the Canadian Pacific Railway and its protective tariff, and settlement through immigration – all of which were central to the development of the Canadian West. Altogether, these government-controlled institutions were meant to establish distinctly Canadian economic and cultural ways of life on the north side of the forty-ninth parallel. Regulating rather than preventing trade across the international boundary, Canadian national policy constantly moderated the economic and cultural reach of modernization. Commodities and practices circulating within the mass market, especially commercial amusements, were often free to move across the border except when policy encouraged overtly nationalizing alternatives.

Friesen pointed to institutions that explicitly facilitated a national policy on the Canadian side of the forty-ninth parallel. Evident in policy, easily isolated, and studied through textual archives, these are without exception repressive state apparatuses in the traditional Althusserian sense: policing, subsidy, tariff, treaty, and settlement (Althusser 1989, 170–86). A potentially more complex question is how the national policy on the Prairies was also instituted through ideological state apparatuses – to stick with the traditional Althusserian scheme – not least because culture largely exceeds the bounds of the official bureaucratic archive and requires methods located in everyday practice. This was, indeed, the subject of Friesen's (2000) subsequent study into the role of communications in national identity. Tracing the circulation of early cinema outlines the tension between national and continental options within the cultural sphere at a time when the border was still largely immaterial in terms of everyday commercial and cultural practice, although firmly established in political and ideological effect. If early film showmen were able to work across or within borderlines, then the case study blurs the distinctions between structural and cultural institutions, beginning with the border itself. If nationalism is the means of establishing acceptance of an imagined community, then the nation's boundary line becomes partly an imagined – rather than simply an institutional – fact that is continually reproduced in daily practice through a variety of state, economic, and cultural apparatuses.

In towns and cities in marginal regions, the local circulation of mass entertainment signified being "in the swim" of modernity, current with metropolitan hubs of commercial and cultural invention and activity. Yet the integrating cultural current of mass culture circulated north-south through the same means of communications and transportation used to support east-west protections and distinctions on either side of the forty-ninth parallel. What follows is a case study of these entertainment routes in the first year of moving pictures from July 1896 into 1897. This moment, when the electric technology of cinema was still a novelty amusement, illustrates the tension at play between established regional routes along the Red River Valley, which treated the border as permeable, and nationalizing networks, which treated it as impermeable. Showmen could either cross the border or respect it, and could promote regional connection or national difference. Big-time American shows began to add cinema to their established cross-border routes; small-time shows were more likely to treat the border as an enclosure. One Canadian showman, Richard A. Hardie, explicitly embraced nationalism to produce Manitoba booster films, intended to promote immigration from the United Kingdom rather than cultural connection to the United States. The technological reproducibility of cinema ironically imposed territorial borders on exhibition circuits, even as it became viable to include small towns as well as villages within those circuits. The long-term impact on the Canadian Prairies was, eventually, to render them marginal to Toronto, part of an Ontario-centric network in a rail-based mass market and yet ever more reliant on technology and content from the United States. Similarly, the American Plains became marginal to Chicago and Minneapolis in national theatre chains. This result was not evident when moving pictures first appeared.

REGIONAL COMMUNICATION AND TRANSPORTATION

A canonical definition introduces mass communication as an extension of trade and transportation. This is especially relevant in Canadian studies following the foundational work of Harold Innis, who explained the economic development of the nation as originating in the fur trade (Berland 2009; Acland and Buxton 1999). A brief summary of trade and transportation in the Red River Valley is worthwhile as a pre-history of cinema in the region. While I risk trivializing a complex history, such attention to the pre-existing means of commercial and cultural circulation is rare in film history, which often proceeds with the interest in cinema defined as a technological invention that was simply imported and unveiled to audiences (Musser 1994; Morris 1992). Partly in order to avoid duplicating the presumptions of its own promotional ballyhoo, I want to highlight how early cinema followed existing communication routes before it forged its own (Gaudreault and Marion 2005).

The "natural" flow of the Red River of the North is inextricable from the global fur trade of the eighteenth and nineteenth centuries, as a central route to transport furs from the northwest territory of British North America and the United States to Europe. As a matter of efficiency, the Hudson's Bay Company supplemented its arduous overland movement of furs down to St Paul, Minnesota, with steamboats on the Red River in 1859 (Gailbraith 1957, 76). Although the prospect of railway expansion into the Dakotas and Manitoba was already in view, the first transcontinental route, the Union Pacific, took until 1869 to complete. Not long after, Manitoba and then British Columbia became provinces of Canada, the latter mandating a railway link between the Canadian Pacific and eastern industry. Before transcontinental routes were completed on either side of the forty-ninth parallel, however, interim railway construction primarily facilitated north-south trade along river paths. The Northern Pacific Railway first reached the Red River at Moorhead in 1871, which added further incentive to develop steamboat transport to Winnipeg. By 1872, the Hudson's Bay Company itself ran the riverboat service in combination with James J. Hill, owner of the St Paul, Minneapolis, and Manitoba Railway. Known at the time simply as the Manitoba Railway, Canadian-born Hill also served on the board of the Canadian Pacific Railway and vied for the obvious economy and efficiency of routing the railway from Winnipeg south along the Red River to St Paul (along his own rails, of course) (den Otter 1983; Martin 1976). Politics intervened and the Canadian Pacific Railway was instead routed entirely within national boundaries north of Lake Superior, becoming a totem of national unity as well as industrial and political independence from the United States (Innis 1971). Even the Canadian Pacific, however, first built a railway link to the international boundary, meeting Hill's Manitoba Railway at Emerson-St Vincent in December 1878. Railway traffic along the

Red River Valley flourished, and the Canadian Pacific opened a second route to Gretna in 1882 while the Northern Pacific opened a line on the western side of the valley from Fargo to Pembina in 1887. Despite these routes, railway development on the whole served national interests with east-west pathways on either side of the forty-ninth parallel: the Northern Pacific Railway from Fargo to Tacoma was completed in 1883; the Canadian Pacific Railway from Montreal to Vancouver was completed in 1886; and Hill's own re-christened Great Northern Railway through Grand Forks met the Northern Pacific at Helena in 1888, before its path to Seattle and north to Vancouver was completed in 1893.[1]

Local performances and community halls in the cities of the Red River Valley came along with settlement, but commercial theatres that hosted touring shows followed in the wake of railway connections and became routine in the 1880s (Hartman 2002; Davis 1989). Substantial theatres opened: the Fargo Opera House in 1883, rebuilt in 1893; the Princess Opera House in Winnipeg in 1883; the Bijou in Winnipeg in 1892, rebuilt as the ornate Winnipeg Theatre in 1897; and Cline's Opera House in Grand Forks in 1884, replaced by the Metropolitan in 1890. While live performance is not usually considered a communication technology, local opera houses operated as nodes for the circulation of itinerant shows. Performed entertainment became a form of modern communication built atop the transportation infrastructure and in continuity with later technologically grounded mass culture and broadcast media. Itinerant entertainment routes followed railway circuits, were promoted through newspapers and telegraphs, produced a consumer experience over a vast territory, and provided metropolitan culture to the hinterland. Travelling shows achieved much the same outcomes as later broadcast technologies, though they worked over a season and toured an adaptation instead of instantaneously reproducing the metropolitan original.

By the mid-1890s, travelling melodramas and variety shows circulated in wide regional markets, including a north-western circuit that began in Chicago and typically toured west through Iowa, north through Minnesota and the Dakotas, across Montana and Washington, and then south through Oregon and Northern California before it headed east again on a more southern path. Notably, this theatrical circuit recognized the entire sparsely populated northwest between Chicago and San Francisco as a common market – within the logistics of railway transportation – that shared

Figure 3.2 Red River Valley theatrical circuit (*Manitoba Free Press*, 13 November 1897)

a common mass culture as a result. Canada was not included on the route except for two key cross-border paths: to the cities of Victoria and Vancouver in British Columbia, and to Winnipeg in Manitoba. Though Winnipeg was by far the most remote and northerly stop, it was worth the additional effort because of its large population. In a sense, this meant that the long-standing cross-border path of the Red River Valley superseded the political and bureaucratic barrier of the international boundary. The Canadian Prairies were not fully incorporated into this circuit until the early twentieth century when two Seattle-based vaudeville circuits, Pantages and Sullivan & Considine, began to operate a second tier of theatres apart from the Chicago-based Orpheum circuit and thus began to exploit new regional markets, including Alberta and Saskatchewan (Butsch 2000).

The dominance of Pantages and Sullivan & Considine over plains and prairies vaudeville illustrates how theatre ownership and management often consolidated regionally on the strength of contracting for big-time touring productions. On a smaller scale, the same happened in the Red River Valley in the 1890s. Starting out with management of the Fargo Opera House, C.P. Walker built a small regional circuit called the Red River Valley Theatrical Circuit, or sometimes the breadbasket circuit (ad for Red River Valley Theatrical Circuit, *Manitoba Free Press*, 13 November 1897). When he opened the rebuilt Bijou as the Winnipeg Theatre in 1897, its joint arrangement with the Fargo Opera House and Metropolitan in Grand Forks was prominently featured in its publicity (*Manitoba Free Press*, 8 May 1897). By 1899, Walker had added Fergus Falls and St Cloud, and opera houses in Crookston and Grafton, for a total of eight theatres from Wahpeton to Winnipeg on either side of the river in North Dakota and Minnesota (*Manitoba Free Press*, 20 May 1899, 22 August 1908; see also Skene 1990). As the most populous city in the region, Winnipeg was a relatively lucrative stop on the continental theatre circuit, and its strongest connection in commerce and communication came through its long-standing connection through river transportation, now railway links, to precisely the set of towns where Walker managed local opera houses.

THE FIRST MOVING PICTURES IN THE REGION

The Red River Valley existed as a cross-border segment within a vast Western US territory for commercial amusements when cinema debuted across the continent throughout 1896. In April, the device known as Thomas Edison's Vitascope became the first moving picture projector in North America available for entrepreneurs to license for public exhibitions. Other moving picture apparatuses were already on the market in France and England, and had been exhibited in limited contexts in New York and, for example, in Minneapolis and at the Atlanta Cotton Exposition in 1895. Edison did not invent the Vitascope – his company merely bought Thomas Armat's Phantoscope and adapted it slightly for manufacturing and marketing (Spehr 2008). The New York marketing firm Raff & Gammon sold Edison's Vitascope to entrepreneurial showmen on a "states rights" basis, with prices that ranged from $6,000 for Massachusetts or Pennsylvania to $500 for the sparsely populated states in the west. International territory, too, was up for grabs, including "the entirety of Canada" for $8,000. Although film historians in Canada have always taken this at face value, in practice the rights to "Canada" covered only eastern provinces. Raff & Gammon offered prospective film showmen separate licenses for British Columbia and Manitoba for $600 each and treated them as distinct territories apart from the rest of the country. The terminology in Raff & Gammon's correspondence is inconsistent, and at one point refers to Winnipeg as a country in its own right.²

North Dakota was one of the first states showmen inquired about for the right to exhibit the Vitascope, outside of the populous northeastern states with lucrative metropolitan cities. John Cryderman, from the relatively remote town of St John just south of the forty-ninth parallel, first wrote to Edison about the Vitascope in March 1896, and was assured by Raff & Gammon that the price for North Dakota was fixed at a low cost of $500, which would ensure a profitable investment. Cryderman emerged as the only serious bidder for the state. He then wrote on behalf of an unnamed business partner about acquiring the rights to Manitoba as well and was quoted a price of $600. Details are missing for the final deal for Manitoba, but local reports show that Cryderman collaborated with Richard A. Hardie on the Manitoba side and that the two shared a projector. Cryderman's entry into the business remains unexplained, since he was not an existing licensee of Edison's Kinetoscope through Raff & Gammon like G.W. Walters in Helena, Montana, or the Holland Brothers in eastern Canada. Raff & Gammon's sales pitch was meant to encourage early investors and licensees to buy further territory, either to exploit themselves or to sub-license to others at a premium once the moving picture craze was underway and their states rights steadily climbed in value. Such value did not arrive. Instead, territorial rights to the Vitascope became liabilities within a few months as competing screen machines that followed an entirely different economic model came to market (Musser 1991, 86–92). Whereas the Vitascope and its films were leased to showmen like Cryderman on top of the territorial rights, independently manufactured competing machines and moving picture films were soon sold directly as a piece of machinery that the owner could exhibit anywhere. Although some competitors violated copyright and patents, machines and films from Europe were available by the end of the summer of 1896. Competing machines all improved on the Vitascope: they were smaller, lighter, and more portable, with films that were clearer, more durable, and projected with better results. Once the craze and competition took off, even Edison Manufacturing undermined its initial support of the Vitascope by selling films to competitors. By the end of 1896, Vitascope showmen had lost their initial advantage of novelty and abandoned their projection outfits. Many begged Raff & Gammon to buy back their territorial rights just to break even.

Competition from independent picture showmen was far from the only problem Vitascope exhibitors faced. Manufacturing delays and electric generation problems routinely led to poor quality or cancelled early picture shows. Raff & Gammon and the Edison factories could not manufacture machines fast enough to deliver the projectors to all of the licensees in a timely fashion, and more remote states rights holders had to wait until showmen in more prestigious, pricier states received their machines. The Vitascope debuted to the public in New York City on 23 April 1896, Boston and Philadelphia in May, and San Francisco and St Louis in June (Musser 2002). Cryderman had initially arranged for his Vitascope to be delivered to Grand Forks in time to debut moving pictures in North Dakota at the Metropolitan during the city's summer fair from 26 June 1896, followed by the Fargo Opera House for the 4 July holiday. If these shows had happened as planned, the Vitascope would have debuted in North Dakota before Chicago, Cleveland, or Des Moines. However, to Cryderman's dismay, the projector did not even arrive in Grand Forks until 6 July and, worse still, it was not compatible with the town's electric supply. The electrical requirements prevented exhibiting the Vitascope anywhere on the North Dakota side of the Red River Valley, and the machine was promptly sent on to Pembina at the Manitoba boundary (*Grand Forks Herald*, 9 July 1896). The North Dakota debut of moving pictures thus occurred on 8 July 1896, with almost no fanfare and only a small report in the *Pembina Pioneer Express* two days later. If Pembina had a different power system than Grand Forks and Fargo, it must have been due to the specifications required for the Canadian Pacific Railway link to town. Certainly the Vitascope's next several stops – all in Manitoba – demonstrate that the cities and small towns of Manitoba had the requisite power supply.

The Vitascope next exhibited in Winnipeg during the city's industrial exhibition, which opened to the public by 18 July 1896 and was promoted in an advertisement in the *Tribune* and a small bit of local news in the *Free Press*. The moving picture show was set up in a vacant Main Street storefront. The next show ran for two weeks in Brandon during the Manitoba Provincial Fair from 27 July, reused the publicity from Winnipeg, and again rented the temporary quarters of a vacant shop downtown rather than appearing at either the opera house or the fairgrounds. Two more Manitoba stops came in August, a few days each in Portage La Prairie and Carberry. In fact, the most detailed and complete local reporting comes from this least populous early host of moving pictures, as the *Carberry News* (14 August 1896) noted with excitement and detail that Hardie's Vitascope was a modern, electric wonders that had appeared in town that week.

In the meantime, Cryderman seemed to give up entirely on the prospect of showing the Vitascope in North Dakota. The only other Vitascope exhibition he or Hardie could possibly have given happened in Duluth, Minnesota, immediately after the machine left Manitoba (*Duluth News-Tribune*, 15 August 1896). It seems Cryderman and Hardie may have simply cut their losses and sent the machine back to New York after only a month trying to entertain audiences. Although Hardie had a successful four weeks playing in four Manitoba cities and towns in return for his $600 states rights, Cryderman's $500 investment gave him only a tiny place in local film history as the first person that managed, against the odds, to exhibit moving pictures in North Dakota.

THE FIRST YEAR OF MOVING PICTURES IN THE REGION

The next several moving picture shows in the region arrived courtesy of small showmen who left no trace except for their names. "Professor" Wymond briefly toured his Magniscope in conjunction with a theatrical stock troupe, the Columbia Opera Company, which visited Winnipeg in December 1896 after playing there for over a month in the previous summer. Meeting up with the company as it crossed back south of the border in Grafton on 15 January 1897, Wymond played moving pictures between the acts. Together, Wymond's Magniscope and the Columbia Opera Company then spent a night in Grand Forks before they moved on to three performances in Fargo the following week. Wymond then toured across the Great Northern railway with the outfit, although there is no mention of moving pictures when the Columbia players visited Devils Lake, their next stop. His name and Magniscope were still part of the show in Washington and Oregon in February and March, but not when it continued to Arizona and Texas later in 1897. How Wymond connected with the company is a mystery, other than the coincidence of being in the same place at the same time.

The next display of moving pictures in the region arrived via Dr Davison's Museum of Anatomy. Grand Forks advertising on 25 February 1897 quoted from positive reviews in newspapers from St Paul, St Cloud, Crookston, and Fargo. Davison's Museum promised a "magnificent collection," the "result of years of study and labor," and to teach "how we live, breathe, and have our being." The show admitted "men only" and was a gruesome, graphic spectacle that combined a human dissection, displays of fifty cases of private diseases, an Eden Musee for "viewing the entry of lost souls into eternal torture," and a Graphoscope moving picture machine with twenty-four "life pictures" (*Grand Forks Herald*, 26 February 1897). The most distinctive part of Davison's Museum, however, was a Phonopticon of projected still pictures of an infamous lynching in Texas in 1895 in combination with a phonograph recording that was purportedly of the same incident. The show stayed in Grand Forks for two weeks, and then made a return trip to Fargo. Although blackface entertainment was a staple in Canada as much as the United States, authorities would have challenged the scandalous, sensationalist, racialized spectacle if it had tried to cross the border to play in Winnipeg.[3] For better or for worse, Davison's anatomical "hellscape" was the second time cinema appeared in Fargo and Grand Forks.

The next appearance of cinema in Grand Forks, a three night show called the Cinematoscope at the end of May 1897, was more typical in hyping its "wonderful living picture machine ... the very latest and best machine and entertainment of the kind on the road" (*Grand Forks Herald*, 23 May 1897). Its publicity cites small-town newspaper reports throughout Michigan from weeks earlier. Although the show came from Racine, Wisconsin, and passed through Brainerd and Fergus Falls, Minnesota, Grand Forks seems to be the farthest west it toured. Another early show that arrived in North Dakota from more eastern states was a special engagement of Kline's Cinematographe during the Devils Lake Chautauqua in early July 1897, a full year after the Vitascope debuted in Pembina.

These small shows all treated the border as a boundary, but moving pictures soon joined major touring shows that had long included Winnipeg on their routes. Moving pictures were thus incorporated

Figure 3.3 Cartoon depicting the men of Grand Forks "studying" the atrocities at the Museum of Anatomy, which included the peepshow of the Eden Musee and images and a sound recording of a Texas lynching, depicted here in the back room (*Grand Forks Herald*, 28 February 1897).

into the performance of Harry Martell's *The South Before the War*, which was billed as "50 People on the Stage, Special Scenery, Novel Specialities, Picaninny Band, and the Wonder Cinematograph." This leg of the long-running show began in Chicago early in August 1897 and headed north through Wisconsin to Duluth and other towns in Minnesota. In the first week of October, it travelled up and down the Red River Valley from Wahpeton to Winnipeg and back to Fargo before proceeding across the Northern Pacific Railway line and arriving in Seattle early in November. As was typical of western US shows, it then toured the Pacific coast from Vancouver to San Francisco before it headed back east through Utah and Colorado. On a similar route, which included Winnipeg along with other cities in the Red River Valley in the first week of November 1897, was an exhibition of the Veriscope films that depicted the famous heavyweight prize-fight between Corbett and Fitzsimmons. Originally filmed in Nevada on 17 March 1897, dozens of copies of the pictures of the fight toured throughout North America for the remainder of the year. Only these wide-ranging tours of major shows crossed the border.

Although independent film showmen in Manitoba also limited their movements to one side of the border, their geographical range and ambition was nonetheless clear. Beginning in May 1897, two competing film showmen, Richard A. Hardie and William McCarthy, traversed the entirety of the Canadian northwest as a single market for entertainment – spanning the head of Lake Superior in northern Ontario all the way to the Rockies and ultimately across the British Columbian interior to the Pacific Coast. Although major US shows included Winnipeg on their western tours in 1897, local showmen could exploit the remainder of the Canadian Prairies. I have already noted Hardie's importance, as he brought the Vitascope to Manitoba in July 1896 after Cryderman had technical troubles with the machine in North Dakota. Hardie had purchased a new Kinetoscope projector of his own in June 1897 and began to exhibit moving pictures at Winnipeg's summer parks. In the Manitoba moving picture field, he now had competition. After Hardie's 1896 Vitascope shows, it took until May 1897 for a second projector to arrive in Winnipeg: an Anamatagraph purchased by William McCarthy of Rat Portage (now Kenora), just on the other side of the Ontario border. McCarthy had no experience in entertainment but was nonetheless versed in showmanship as an exploiter of mining claims and other ventures near Rat Portage. McCarthy arranged to partner with the Cosgrove Comedy Company, a well-known variety show of John Cosgrove and family from St Mary's, Ontario. The company had extensively toured the Canadian Prairies several times since 1891, and was headed westward again through northern Ontario just as McCarthy bought his projector. The Anamatagraph debuted with a two-night engagement at the Port Arthur Town Hall on 10 and 11 May 1897, then to neighbouring Fort William and onto Rat Portage. After a week-long engagement in Winnipeg for Victoria Day in May, the Anamatagraph and Cosgrove Company toured along the southern railway routes of Manitoba and visited towns that included Carmen to Wawanesa, Souris to Morden, and a dozen others.[4] When it returned to Winnipeg mid-July, the entire business arrangement ended. Cosgrove partnered with Hardie's Kinetoscope. Meanwhile, McCarthy assembled his own troupe and created the Royal Anamatagraph and Speciality Company. Early in August, McCarthy set out with the Anamatagraph across the main Canadian Pacific Railway line, having the advantage of being the first to exhibit moving pictures across the northwest territories in the present-day provinces of Saskatchewan and Alberta. The show proceeded as far west as Macleod and Cardston in Alberta without venturing to Calgary and Edmonton, and the return leg of the tour solicited only meagre attendance and an audience that was disappointed that the pictures were the same as on the previous visit.

If repeated views of moving pictures were already a problem, a new set of films came to Assiniboia and Alberta soon enough. Hardie's Kinetoscope joined the Cosgrove Company for its own tour westward across the Prairies, beginning in Brandon and traveling to points north of the main CPR line in Manitoba in August 1897. In Winnipeg in early September, Hardie pursued his latest venture: local moving pictures. Only a few of the pictures are mentioned specifically in publicity: the Winnipeg and Brandon fire brigades racing down city streets, sidewalk crowds, trains racing toward the camera, and plenty of wheat being harvested, including Manitoba Premier Thomas Greenway at work in his own fields. These are the same films James S. Freer brought to England in 1898 as an immigration and settlement promotional tool (Morris 1992, 30–3). Although Freer has long been assumed the filmmaker, he did not travel aboard until December 1897, months after the films had already been exhibited across the Prairies. They first appeared in Winnipeg in September 1897 for officials from the railways and government (*Manitoba Free Press*, 11 September 1897; 9 December 1897). In September and October of 1897, Hardie and the Cosgroves toured the path taken by McCarthy's Anamatagraph. They stopped at more towns on more remote railway spurs and brought moving pictures to Prince Albert, Edmonton, and Calgary, with a rest in Banff before they returned to Manitoba. In November and December, the company began an extensive tour of the southern part of Manitoba and appeared in even smaller villages than before, often spending just one night in each before moving onto the next.

At the beginning of 1898, just as Freer headed to England, Hardie and Cosgrove severed their partnership and set off individually on yet another tour across the Prairies – this time in winter – each with his own brand new Kinetoscope, copies of the Manitoba films, variety acts in support, and disparaging remarks about how his rival was a mere copy. Hardie headed straight out via the CPR line, while the Cosgrove Company began in the northern part of Manitoba and travelled north to Yorkton in present-day Saskatchewan. Cosgrove's engagements were uniformly shorter by a day or more, and the rivals recklessly began to play the same towns just days apart. They played in Edmonton simultaneously, to the amusement of the local press and the confusion of the public. Hardie returned to Winnipeg in March 1898, but Cosgrove travelled with his Kinetoscope over the Rocky Mountains and brought moving pictures to Kamloops and Kelowna for the first time. An American showman had brought the first moving pictures to British Columbia in Victoria in February 1897, but Cosgrove's Kinetoscope was the first to arrive on the coast via the Canadian Pacific Railway.

Taken as a whole, the first year of cinema in the Red River Valley borderland region relied on a confluence of contradictory political, economic, and technological mediations of the international boundary. The new amusement could be facilitated by cross-border commercial ties, but was also used to make national distinctions in how people related to the new form of communication. The Red River of the North runs perpendicular to the international boundary of the forty-ninth parallel, which created a stark choice for showmen. If they ignored the border, they could follow the flow of the river and seek out a naturalized route between large populations gathered in bigger cities, a route charted by cross-border railway lines built to follow the earliest river-based trading routes. This path was primarily followed by the big-time entertainment of Chicago-based professional shows. However, the river path was also followed by the international partnership between Cryderman and Hardie, who separately bought Edison's Vitascope rights to North Dakota and Manitoba, respectively. They ended up sharing the same projector and films, either by plan or by circumstance when the machine did not work in Grand Forks.

The alternate route respected the border and did not cross. Small-time regional showmen exploited their markets more thoroughly than the big-time shows, but it was their overtly nationalist entertainments, rather than scant resources, that turned the border into a barrier. For example, Dr Davison did not bring his Museum of Anatomy to Winnipeg, despite the lucrative marketplace just a few hours train ride away. This was less a matter of cost than avoiding the chances of customs agents confiscating his perhaps-obscene material. His sensationalist entertainment was too overtly American in its emphasis on racial violence during a time when progressive and moralistic arguments were just crystallizing in Canada against the pernicious influence of American popular culture (Gabriele 2011). Thus, Davison and other small-time American shows used only the east-west railway routes rather than those that crossed the border.

Richard Hardie's Manitoba farming films were also part of a nationalist project, and so he did not bring them to the agricultural towns and cities in the heartland of the Dakotas on the southern part of the Red River Valley. His intention to use the films to promote immigration to Canada, not to mention his sponsorship by Canadian railways and governments, superseded any sense that the farming scenes would be of interest to American farmers in a farming region. Instead of heading south along the Red River Valley pathways, he and other small-time Canadian showmen used the nation-building institution of the Canadian Pacific Railway to bring their amusements to the Canadian Northwest Territories, eventually to the British Columbia coast. Their entertainment, in a very different fashion than Davison's sensational anatomical museum, were equally part of a nation-building project.

Entertainment and mass culture sometimes sit atop the railway and other nation-building institutions to reinforce the border across the forty-ninth parallel, and sometimes treat it as irrelevant. Aside from reporting a history of film in a region relatively neglected in film history, the decision to isolate the Red River Valley as a cultural region in the first place was only indirectly due to its being a geographic watershed. The river determined a regional boundary because cultural industries at the turn of the twentieth century still largely circulated through transportation routes that originated in the river itself. Alternative national routes had just been forged, which allowed an alternative nationalist orientation to cultural communication. Tracing entertainment routes thus displays the tensions at play in mediations of mass culture and communications in this borderland, and across the border in general.

NOTES

1 Henry V. Poor compiled an annual survey of railways in North America in his 1868 *Manual of the Railroads of the United States*.

2 All references to Raff & Gammon correspondence are from letterbooks archived at the Baker Library of the Harvard Business School, MSS 692, volume 2 (outgoing letters April to May 1896), volume 3 (outgoing letters 1895 to March 1896), and volume 6 (incoming letters, especially file 24 from Cryderman in North Dakota).
3 For one case study of blackface in Canada, see Nicks and Sloniowski (2010).
4 All information about the tours of McCarthy's Anamatagraph and Hardie's Kinetoscope comes from notes in local newspapers. See Moore (2012).

REFERENCES

Acland, Charles R., and William Buxton, eds. 1999. *Harold Innis in the New Century: Reflections and Refractions*. Montreal: McGill-Queen's University Press.

Althusser, Louis. 1989. *Lenin and Philosophy and Other Essays*. London: New Left Books.

Berland, Jody. 2009. *North of Empire: Essays on the Cultural Technologies of Space*. Durham, NC: Duke University Press.

Bumsted, J.M. 2008. *Lord Selkirk: A Life*. Winnipeg: University of Manitoba Press.

Butsch, Richard. 2000. *The Making of American Audiences from Stage to Television, 1750–1990*. New York: Cambridge University Press.

Davis, Ronald L. 1989. "Opera Houses in Kansas, Nebraska, and the Dakotas, 1870–1920." *Great Plains Quarterly* 9: 13–26.

den Otter, A.A. 1983. "The Hudson's Bay Company's Prairie Transportation Problem, 1870–85." In *The Developing West: Essays on Canadian History*, edited by John E. Foster, 25–48. Edmonton: University of Alberta Press.

Friesen, Gerald. 1987. *The Canadian Prairies: A History*. Toronto: University of Toronto Press.

– 2000. *Citizens and Nation: An Essay on History, Communication and Canada*. Toronto: University of Toronto Press.

Gabriele, Sandra. 2011. "Cross-Border Transgressions: The American Sunday Newspaper, the Lord's Alliance and the Reading Public, 1890–1916." *Topia* 25: 115–32.

Galbraith, John S. 1957. *The Hudson's Bay Company as an Imperial Factor, 1821–1869*. Los Angeles: University of California Press.

Gaudreault, André, and Philippe Marion. 2005. "A Medium is Always Born Twice ..." *Early Popular Visual Culture* 3 (1): 3–15.
Hartman, James B. 2002. "On Stage: Theatre and Theatres in Early Winnipeg." *Manitoba History* 43: 15–24.
Innis, Harold. 1971. *A History of the Canadian Pacific Railway*. Toronto: University of Toronto Press.
Martin, Albro. 1976. *James J. Hill and the Opening of the Northwest*. New York: Oxford University Press.
Moore, Paul S. 2012. "Mapping the Mass Circulation of Early Cinema: Film Debuts Coast-to-Coast in Canada in 1896 and 1897." *Canadian Journal of Film Studies* 21 (1): 58–80.
Morris, Peter. 1992. *Embattled Shadows: A History of Canadian Cinema, 1895–1939*. Montreal: McGill-Queen's University Press.
Musser, Charles. 1991. *Before the Nickelodeon: Edwin S. Porter and the Edison Manufacturing Company*. Los Angeles: University of California Press.
– 1994. *The Emergence of Cinema: The American Screen to 1907*. Los Angeles: University of California Press.
– 2002. "Introducing Cinema to the American Public: The Vitascope in the United States, 1896–97." In *Moviegoing in America: A Sourcebook in the History of Film Exhibition*, edited by Gregory Waller, 13–26. Malden, MA: Blackwell.
Nicks, Joan, and Jeannette Sloniowski. 2010. "Entertaining Niagara Falls: Minstrel Shows, Theatres, and Popular Pleasures." In *Covering Niagara: Studies in Local Popular Culture*, edited by Joan Nicks and Barry Keith Grant, 285–310. Waterloo, ON: Wilfrid Laurier University Press.
Skene, Reg. 1990. "C.P. Walker and the Business of Theatre: Merchandising Entertainment in a Continental Context." In *The Political Economy of Manitoba*, edited by James Silver and Jeremy Hull, 128–50. Regina: Canadian Plains Research Centre.
Spehr, Paul C. 2008. *The Man Who Made Movies: W.K.L. Dickson*. Eastleigh, UK: Libbey.
Stuart, Reginald C., and M. Brook Taylor. 2005. "The Epic of Greater North America: Themes and Periodization in North American History." In *New England and the Maritime Provinces: Connections and Comparisons*, edited by Stephen J. Hornsby and John G. Reid, 280–94. Montreal: McGill-Queen's University Press.
Thompson, John Herd, and Stephen J. Randall. 2008. *Canada and the United States: Ambivalent Allies*. Fourth edition. Athens: University of Georgia Press.

THE POLITICAL BORDER

4

"Shutting Down the Snake Ranch": Battling Booze at the BC Border, 1910–14

BRANDON DIMMEL

For many residents of the Pacific Northwest, the Peace Arch is the defining symbol of the Canada-US international boundary. Unveiled in 1921 to commemorate a century of peace between the two countries (1814–1914), the Peace Arch is a towering sixty-seven feet of concrete and reinforced steel. To emphasize this bold idea of a peacefully shared continent, inscriptions on the sides read "Children of a Common Mother" and "Brethren Dwelling Together in Unity," proclamations about Americans and Canadians that, when originally unveiled, fed on and into strong feelings of Anglo-Saxon brotherhood in the period following the Allies' victory over the Germans in 1918. Almost a century later, the Peace Arch today is surrounded by a carefully manicured international park, a beautiful commons that every year attracts thousands of visitors who, because it is flanked by Canadian and American customs stations that connect Interstate 5 with BC Highway 99, mostly take in the colourful display from their cars.

But things have not always been so peaceful at this idyllic place between the seaside towns of White Rock, British Columbia, and Blaine, Washington. In fact, the location was once at the heart of an intense and bitter debate between residents of the nearby communities. Shortly before the First World War, controversy erupted over a single establishment: the St Leonard Hotel, which served whiskey and beer just a stone's throw from the international boundary and mere feet from where the Peace Arch stands today. A two-storey, ten-room, white, wooden structure of considerable girth, it was referred

Figure 4.1 Peace Arch unveiling, 6 September 1921. The St Leonard Hotel stands to the left (White Rock Museum and Archives, image 1996-36-107).

to in Blaine, which opted for local prohibition in 1910, as the "hellhole on the border" (Ellenwood 2004, 143). In White Rock, locals had mixed feelings about the hotel, some defending it because owner and proprietor Richard Asbeck ran a Canadian establishment breaking no Canadian law. In this paper, I contend that between the time Blaine went dry in 1910 and the St Leonard Hotel closed in early 1914, the debate that surrounded the saloon demonstrated to residents of Blaine and White Rock that the border, which was not marked by a natural or man-made barrier, did exist, and that it separated people of different beliefs with regards to social reform. The fact that the St Leonard Hotel stood mere feet from the border helped to emphasize this separation, as well as the idea that the international boundary was a metaphysical entity that needed to be observed, defended, and maintained. This marked an important shift for a region where trans-boundary movement had once been a necessary part of surviving the untamed Pacific frontier.

INTERPRETING AN INVISIBLE BOUNDARY

Around the turn of the twentieth century, the border's role in North American economics and culture was a sensitive topic for many Canadians. In Ottawa, the federal government struggled to stabilize an often-chaotic economy based on the export of primary resources like fish, timber, and agricultural goods. In 1879, Conservative prime minister John A. Macdonald decided the best course of action was protectionism and placed high tariff walls between Canada and its neighbour, which many Canadians – particularly businessmen from the country's manufacturing centre in Ontario and Quebec – felt could easily overrun the economy with cheaper goods produced in New England or the rapidly developing industrial belt wrapped around the Great Lakes (Smith 1891, 207). But not everyone agreed with this policy. Besides the Liberal Party, which advocated lower tariffs and conditional reciprocity between the United States and Canada, influential economic and political experts carefully devised their own theories on how the boundary should affect relations between the two nations. Arguably, none of these experts had more influence than Goldwin Smith, a Briton educated at Oxford, once employed at Cornell, and later a political commentator based in Toronto, who in his widely read book *Canada and the Canadian Question* (1891) argued that the Canadian nation was a political and cultural absurdity. He believed the geographic borders that separated each Canadian region were more substantial in shaping the economic and cultural outlook of the country's approximately five million residents than the invisible forty-ninth parallel or waterways like the Detroit and Niagara Rivers (1891, 1). According to Smith, a resident of the mountainous, densely forested British Columbia had far more in common with the salmon fishers and lumbermen of Washington or Oregon – with whom he might actually have a conversation – than with the distant factory workers and office clerks of Toronto or Montreal.

However, the idea that the Canadian nation and its border was little more than an absurdity upset many Canadians, particularly those who were British-born or of British heritage, many of whom emphasized the country's Anglicized culture and economic ties as a critical distinction from the United States. In the period after the First World War, historians like Arthur Lower and artists like the Group of Seven pointed, generally speaking, to Canada's climate and geography as a way to differentiate between a Canuck and a Yank (Berger 1995, 112; Hill 1995, 16).

Unfortunately, few of these studies examined how people who actually lived near the border felt about the boundary or differences and similarities between themselves and nearby Americans. Could, as Smith suggests, one's geography and industry (fishing versus farming or factory work) really determine the extent of one's nationalism? Could, as his critics have since posited, the climate and Anglicized heritage have helped Canadians distinguish themselves from Americans, even those just a few miles away? Perhaps. But there are other ways to understand how residents of Canada and the United States felt about national identity and the international boundary in the past. For one, it is possible to examine national and regional identities by examining individual, even small, communities located at or near the border.

PEACEFUL BEGINNINGS, PERMEABLE BORDER

The St Leonard Hotel affair betrayed a history of amicable relations between White Rock, the surrounding British Columbia county of Surrey, and Blaine. For years prior, this had been a crossing worthy of a Peace Arch: prior to the turn of the century, the only general store in the region was located on Semiahmoo Spit, a short extension of land that juts out into Semiahmoo Bay from downtown Blaine. In the late nineteenth century, Surrey pioneers regularly walked to Blaine and then paddled rowboats to the store to purchase tools and foodstuffs (Stewart n.d.). Those staffing the first customs station in the region, established at Elgin just north of present-day White Rock, recognized the store's importance to locals and permitted Canadian settlers to declare goods purchased in Blaine once every six months rather than after every crossing (Ellenwood 2004, 134). White Rock's settlement began in the 1880s and owes much to a Blaine businessman known today only as Gerbritsh, who subdivided the townsite and sold lots to residents of south Surrey and his own home town (*Semiahmoo Gazette*, 1 August 1916). Throughout the pioneer period and indeed into the 1950s, White Rock lacked an effective fire protection agency and relied on Blaine to bail residents out of trouble during the long, dry summer months (*Semiahmoo Sounder*, September 1991; Ellenwood 2004, 113). For their part, Blaine residents often crossed into White Rock to relax on the beach or to cut timber

(Ellenwood 2004, 68).¹ For the majority of the late nineteenth century, this was a peaceful boundary in a still-wild Pacific Northwest.

The St Leonard Hotel, named after a similar establishment in Sussex, England, was built in 1883 by British immigrant Bailey Ross (*Surrey Now*, 12 March 2010). Ross had high hopes for his saloon as immigrants began to move into both Blaine and, to a lesser extent, White Rock in the years that followed. Unfortunately, progress was slow and fortunes were delayed. After a promising start with the building of its first salmon cannery in 1875 and other new businesses like a shingle mill and sawmill in the early 1880s, Blaine's momentum died by the end of the decade. The Elwood sawmill moved away and the shingle mill burned down. A proposed railway connection between Blaine and the Northern Pacific line, which would have created work for dozens of local men, was abandoned due to recession (*Blaine Journal*, 30 July 1959). When Blaine residents fell on hard times, the municipality of Surrey across the line felt the impact, as desperate Americans hunted Canadian grouse and trout to the point of exhaustion (Stewart n.d.).²

Progress in White Rock was even slower. In 1890, a Blaine realtor subdivided the land there and began to sell lots for between $150 and $300. New Westminster, British Columbia, businessman John Hendry helped fund the construction of a hotel along with a rudimentary pier, but in the time before efficient communication links – there was no railway connection, nor were there many roads – the ventures collapsed (Ellenwood 2004, 41). Although roads steadily improved over the next two decades and buildings rose on some of those original lots, White Rock was a very quiet place until 1909, when the St Paul, Minnesota-based Great Northern Railway rerouted its connection from the heart of Surrey to the more evenly graded path around the edge of Semiahmoo Bay, directly through the seaside community. From that point forward, White Rock's development as a resort area gained traction. In the next five years its population steadily increased in size, from a few dozen families in 1910 to a year-round population of 430 in 1913. Most important was its vibrant summertime activity, as 3,500 visitors crowded into local shops, restaurants, hotels, and beaches (*Semiahmoo Gazette*, 5 February 1914).

The Great Northern Railway connection benefitted Blaine as well, but local business leaders focused on the fishing and lumber industries rather than on tourism in the years that followed (Arbuckle et al. 1984, 16). Social prospects in the backwoods working-class town were few, and many men chose to spend their leisure hours swilling brew at the St Leonard Hotel just across the line in Surrey, British Columbia. For a time, they were kept company by a number of prostitutes who inhabited a cluster of shacks that surrounded the hotel (Stewart n.d.). Although Blaine conservatives grumbled about the free-flowing booze and debauchery for decades, not until 1910 did the community's growing number of temperance advocates successfully pass local prohibition and ban the sale or consumption of alcohol in town.

Much to their frustration, Blaine's social reformers had no power over the operation of the St Leonard or Surrey Council, which had the final say on the hotel's licensing status. Throughout the period from 1910 to 1913, the bone-dry American border community made little headway in convincing the majority of Surrey residents to shut down the saloon. Certainly, many Canadians in the area recognized that the St Leonard Hotel was a dangerous and immoral place. In an oral interview on Surrey's late nineteenth century development, long-time resident Margaret Stewart states that the establishment had "an awful bad name," and that there "was drinkin' and all kinds of corruption around it." She tells of how her father, who first homesteaded in nearby Hall's Prairie, fought to keep it from being built in the first place: "My father did everything to keep it from being built there because he knew it would be just a dance hall on the [boundary] line" (Stewart n.d.). Still, the Stewart family was outmatched by a much larger contingent of Surrey "wets" and local borderland residents who exhibited a rather apathetic attitude towards Blaine's experiment with prohibition.

There was no single reason for Surrey's refusal to close the saloon. Part of the issue may have been financial: prior to the First World War, the largely agricultural municipality was not particularly wealthy, and the $400 the St Leonard paid in licensing fees each year must have been very difficult for local politicians to turn away (City of Surrey, 21 June 1913).[3] The idea of banning one bar near the international boundary might have opened the door to further closures of other licensed establishments, lowering the municipality's income and slowing the development of its infrastructure. Indeed, this idea of a border region no-booze zone eventually became a part of Blaine's campaign to close the St Leonard.

Figure 4.2 The St Leonard was situated so close to the border that this photograph caption incorrectly lists Blaine, WA, as the hotel's location (White Rock Museum and Archives, image 1995-13-183).

IN SICKNESS AND IN HEALTH: CHALLENGES TO A CROSS-BORDER RELATIONSHIP

For three years between 1910 and 1913, tensions associated with the St Leonard Hotel grew steadily more intense. The issue finally came to a head in early 1913 when the Surrey Medical Health Officer, Dr F.D. Sinclair, attributed a spreading diphtheria outbreak to Blaine's salmon canneries' dumping of animal waste into Semiahmoo Bay. Washing ashore in White Rock only a few miles away, the offal, Sinclair posited, had caused an initial diphtheria case that afflicted a Vancouver visitor to spread quickly among residents of the Canadian border town (City of Surrey, 6 January 1913). At the time of Sinclair's initial report in January 1913, four cases were reported, one ending in death. Unfortunately, this was not to be the end of the dilemma: the next month, Sinclair reported a new case to Surrey Council, which presumably afflicted a child since he noted that the White Rock one-room school house had been shut down to prevent exposure to the students and teacher (City of Surrey, 15 February 1913).

The diphtheria outbreak was more than a mere nuisance for White Rock: instead, it threatened the town's development as a popular seaside resort community. Its clean, warm air, sandy beaches, and panoramic view highlighted by the Olympic mountain range and towering Mount Baker led locals and visitors alike to draw comparisons with the Bay of Naples, giving the community an almost exotic feel. By March 1913, four trains made their way between White Rock, New Westminster, and Vancouver, leaving the latter at 12:15 a.m., 9:30 a.m., 12:15 p.m., and 4 p.m. each day (*Semiahmoo Gazette*, March 1913). Starting in 1911, a "Campers Special" train offered lower rates and special weekend fares for visitors between June and October (Ellenwood 2004, 50). The rates were so low and the service so popular that the Great Northern Railway, which profited little from the venture, unsuccessfully tried to shut down the Special several times.[4] White Rock's popularity as a resort community that attracted visitors from across British Columbia and Washington had grown at a rapid pace since the rerouting of the Great Northern Railway in 1909, and the sight of salmon offal, let alone its consequences for the health of visitors and permanent residents, was troublesome for community leaders who recognized that the boom could go bust if Blaine continued to dump into Semiahmoo Bay.

As part of their response to the diphtheria crisis, White Rock residents pressed Surrey Council to have the issue addressed by a higher authority. The sanitation question made its way to the provincial government and the International Joint Commission, a binational body established in 1909 to help prevent and resolve disputes involving Canada-US boundary waters (City of Surrey, 6 January 1918). In the meantime, locals encouraged Blaine residents to address the matter. It was a request that unleashed three years of pent-up frustration over the St Leonard Hotel's operations. In a heated response from the American town's newspaper, the *Blaine Journal*, titled "Surrey has a Kick Coming," owner and editor J.W. Sheets requested the International Joint Commission first deal with a more troublesome issue: Surrey liquor. "The people on this side of the international boundary have also a job for the International Joint Commission, since it has become known that there is such a body," Sheets wrote. "Instead of rotten fish however, the people over here have a kick coming on rotten Canadian booze, and instead of being a menace to 1,000 people during the summer it is a menace to four or five thousand people all the year" (13 January 1913). Sheets hesitated to blame the "majority" of Surrey's residents, which he felt were opposed to the saloon but were powerless against the "whiskey and brewery interests of British Columbia." Still, he was more than willing to sacrifice their health in order to have the saloon closed: "They want co-operation from this side in remedying their nuisance, but have for three years absolutely ignored complaints arising from this side as to their back door saloon ... This paper asks the people of Surrey to lend their aid in blotting out the shame of British Columbia and in return they will have every possible assistance in dealing with the menace to health complained of."

As tensions continued to rise, the St Leonard Hotel issue drew the attention of Washington State authorities. In late February 1913, Senator Ed Brown introduced and had passed in the state senate a memorial requesting that the provincial authorities at Victoria take action to close the St Leonard Hotel and prohibit the sale of alcohol within two miles of the international boundary. It was a measure that threatened not only the saloon in question, but also similar establishments in BC border communities like White Rock or Huntingdon (*Blaine Journal*, 21 February 1913). In April, Surrey Council received a letter on behalf of British Columbia's lieutenant governor that forwarded messages by British Ambassador at Washington Sir Cecil Spring Rice and Washington Secretary of State William Jennings Bryan, who complained about "the illegal sale of liquor in a hotel in Surrey Municipality situated near Blaine" (City of Surrey, 12 April 1913). Those lofty offices requested the council launch an investigation into the hotel's practices immediately in order "to improve conditions" between Blaine and its Canadian neighbours. As frustrations continued to rise into the fall of 1913, Washington Senator Wesley Jones angrily referred to the St Leonard as a "grog-shop" established solely to prey on Blaine men and exploit the American town's prohibition law (Moore 2006, 252).

Surrey Council's response, as had been the case throughout the previous three years, was that the matter had been investigated and that the council "could find no breach of the law in connection with the matter" (City of Surrey, 12 April 1913). It had the same reply for the Grand Council Royal Templars of Temperance, who called the hotel "disgraceful" and encouraged "some remedy" to the situation soon. In every case of complaint up until 1914, Surrey Council responded that there was no clear indication the St Leonard was operating outside of the law, but the reasoning behind this position may have transcended legal principles. The council was likely impressed with owner Asbeck's attempts to improve the place, given that since taking ownership in November 1910 he had added two flush toilets, water connections, a bath, electric lights, a telephone, a new stable, wagon and poultry sheds, and grass. He had also repainted at least once (City of Surrey, 12 April 1913). Given that most White Rock residencies and much of Surrey lacked telephone connections or electricity at the time, it might have seemed a shame to shut down a business that had shown considerable initiative. The fact was that the hotel was profitable, it paid the council substantial liquor licensing fees, and it was not breaking Canadian law.

Such intransigence continued to anger the people of Blaine, who made it clear that the St Leonard Hotel was more than a mere nuisance: instead, it was a dangerous place that robbed local men of their money and dignity, threatened to arouse public violence and insurrection, and, most importantly, was an affront by Surrey and its council. The *Blaine Journal* closely followed the hotel's role in local crime over the year 1913, and by summer's end the establishment had accrued quite the rap sheet. On 7 February, the *Journal* reported that a Blaine man had been robbed of $158 after being beaten by St Leonard drunks. Two months later, three Whatcom county men were arrested in Blaine for giving away liquor secured from the hotel (*Blaine Journal*, 11 April 1913). In May, another man was caught with St Leonard booze, only his was not a one-time affair. Nineteen-year-old Dolman "Pinkey" Dell admitted to authorities that he had smuggled as many as one hundred bottles between the St Leonard and Blaine before they caught him (*Blaine Journal*, 16 May 1913).

Perhaps the most dangerous situation involving the St Leonard emerged in July 1913 after two Blaine men, Perry Graham and David Reid, were arrested for drunkenness. In an attempt to set an example for other residents, authorities in Blaine placed a $200 bond on the men with plans to transfer both to Whatcom county capital Bellingham for an appearance before a judge – a much sterner approach than the $45 fine usually placed on drunks (*Blaine Journal*, 11 July 1913). In the following days, friends of Graham and Reid led by John "Jim" Hamlin, a "British Columbian inflamed by booze" imbibed at the St Leonard, attempted a jailbreak prior to the transfer. Luckily for the people of Blaine, the local marshal prevented a riot by pre-emptive strike, arresting Jim Hamlin and immediately transferring him to Bellingham. The *Blaine Journal* noted: "That the plans of the gang were laid over the bar at the St Leonard booze joint is now well established. And several things point to the belief that the idea was started by parties in this city who saw that plenty of liquor was furnished to get things started" (11 July 1913).

Blaine's frustration with the St Leonard revealed changes in the way locals viewed the international boundary with Surrey. Gone were the days when White Rock and area residents were welcome to walk freely across the line. Now, the *Blaine Journal* called for more federal agents at the boundary. This was demonstrated in a *Journal* editorial just a week after the Graham and Reid jailbreak fiasco. The occasion: the July 1913 opening of the Pacific Highway customs station on Blaine's east side, which introduced a new crossing that would eventually supplant the St Leonard location as the region's busiest. The trouble for the *Blaine Journal* was that the majority of the town's US customs inspectors were transferred to the new station, which left the area at the St Leonard Hotel, directly across from downtown Blaine, less well patrolled. The *Journal* estimated that upwards of one hundred vehicles, let alone foot travellers, crossed at the station adjacent to the St Leonard each and every day. Even with a full roster of border agents serving there, "a great many [travellers] fail to report after crossing," it noted. The editorial placed little faith in the ability of Blaine police officers or Canadian and US boundary inspectors at the Pacific Highway station, downtown White Rock, and the St Leonard location to keep the violence and rum running from cascading out of control: "It is hoped that this station [at the St Leonard] will not long be left without attention, a station that is today next to the Great Northern railway, the most traveled along the border for many miles" (18 July 1913).

For the next few months, tensions aroused by the St Leonard and salmon cannery issues continued to build among White Rock and Blaine residents. The levee broke in September 1913, when residents of Surrey, led by a White Rock contingent frustrated with the continued appearance of salmon waste on their otherwise pristine shores, began to press Surrey Council to take action against Asbeck and the hotel. The group presented the council with a petition signed by 130 area residents demanding the St Leonard's license be cancelled because of "instances of law violation on the part of the management." However, they too received the council's typical response up to that point: without a conviction, it could not terminate the liquor licence. Despite the setback, the *Blaine Journal* was encouraged by the petitioners' efforts, offering them wry commendations for "able work in behalf of cleaning up their back yard" (12 September 1913).

Rather than convincing Blaine residents to consider alternatives to polluting Semiahmoo Bay, Surrey residents' efforts to shut down the St Leonard only encouraged the continuance of a blackmailing strategy among the American town's anti-liquor interests. On 19 September 1913, the *Journal* noted: "Surrey municipality, we understand, will ask the officials of Blaine to take action against the dumping of fish offal into the waters of the bay on the ground that the practice is endangering the health of the residents in White Rock. This paper sincerely hopes that the city council will courteously decline to make any move until the Surrey officials make a move toward cleaning up their snake ranch on the border here." The *Journal's* editor remained convinced that rotten fish was the "lesser evil" next to the booze joint, and concluded that "public sentiment over there can close this snake ranch and when it does the people of Blaine will meet their neighbours half way and deal with the rotten fish nuisance."

After another violent episode was blamed on the St Leonard in October, big city newspapers began to comment on events unfolding along the border (*Blaine Journal*, 17 October 1913). The most significant coverage came courtesy of Vancouver's *Saturday Sunset*, whose judgment of the situation fell definitively in favour of Blaine's prohibitionists and not Surrey Council or the St Leonard Hotel: "Blaine, be it known, has been a 'dry' town for three years. Those who have visited the city declare it is painfully dry. Thirty feet from the border, and not ten minutes walk from almost any part of the town of Blaine, stands the St Leonard Hotel, affectionately described by the good people of Blaine as the 'hell-hole'" (15 October 1913). Despite attempts by Surrey residents to have the hotel's license revoked, so angry were Blaine residents with their neighbours, the *Sunset* reported, that they refused to step foot across the border to protest the issue with Surrey Council: "That is how matters stand today. Such an occurrence in Canada seems hardly possible, yet it is a fact. All the efforts of the people of Blaine, whether their view is right or wrong, are defeated by the one hotel just across the line. This seems hardly fair."

The *Blaine Journal* noticed that White Rock's *Semiahmoo Gazette* had reprinted the *Sunset* piece and immediately jumped to the conclusion that most residents of the nearby Canadian town were on board with the new campaign to have the St Leonard closed: "The last issue of the *Semiahmoo Gazette* of White Rock copied the article in full along with other matters which clearly showed its disgust of this 'hell-hole' on the border ... It is evident that public opinion is aroused over this great injustice and that it is only a matter of a short time until the joint will be closed" (24 October 1913). Shortly thereafter, the *Gazette* shot back that there must have been a misunderstanding and called the issue a "difficult matter" that was made complex because Asbeck was, in the words of editor Charles Sands, "a worthy man" who was "earning his living honestly, according to his idea and according to [the] law of Canada" (8 November 1913).

The real issue, Sands emphasized, was Blaine's salmon canneries dumping fish offal into Semiahmoo Bay. "Blaine has a 'stink-hole' of her own, which, year after year and season after season polutes [*sic*] and defiles Canadian waters," Sands wrote, adding that the difference between St Leonard drinking and salmon cannery dumping was clear: Blaine men made a choice to cross the international boundary to wet their whistles at the "hell hole on the border," but residents of the Canadian town had no say in when and where fish offal would wash ashore, sending local residents and visiting campers scrambling and hurting White Rock's chances of becoming a popular resort location. Furthermore, as Surrey Council had repeatedly emphasized, the St Leonard Hotel was not breaking any law, while Blaine, in allowing its canneries to pollute boundary waters, "was not fulfilling the tenets of international law" (*Blaine Journal*, 8 November 1913). Sands remained convinced that Blaine had caused the standoff over the St Leonard and it was up to the American town's leadership "to see that this matter is put right."

The St Leonard Hotel issue was at the heart of both the Blaine city election in 1913 and the Surrey municipal campaign in early 1914. In Blaine, the *Journal* called for a leadership that would be fierce in stamping out the liquor traffic from BC and in prosecuting anyone who returned from the St Leonard with an extra stumble in their step. Editor J.W. Sheets requested voters oust the present council, led by Mayor Willison, and instead elect councillor Andrew Danielson, who in the year prior had played an integral role in replacing the city's allegedly corrupt chief of police (*Blaine Journal*, 7 November 1913). Equally revealing of the St Leonard's impact on local politics was the second major issue in the election: voting by former residents of Blaine now living in British Columbia. "Are residents of British Columbia to be allowed to come over here and 'but[t] in' on our elections?" asked the *Journal* in November 1913. The case of Julius Wolten, who after living in Vancouver for two years requested a vote in Blaine's fall 1913 election, prompted the debate. To the frustration of the *Journal*, Blaine's city clerk ruled that former resident Wolten was entitled to cast his ballot because he had kept his US citizenship (7 November 1913, 14 November 1913). "If such persons have the nerve to ask for a vote, they ought to be compelled to defend their rights in court," Sheets shot back, adding, "such persons owe at least to show good intentions by residing here a few weeks before asking to be registered" (*Blaine Journal*, 7 November 1913). Only increasing the friction between Blaine and its Canadian neighbours was an investigation by the Vancouver Board of Trade into the acquisition of tiny Washington territory Point Roberts (accessible by land only through BC). "The proposition has met with more or less ridicule on this side of the boundary line, and very likely will never be considered seriously by our government," Sheets snapped (21 November 1913).

Across the border, popular White Rock businessman Henry T. Thrift – whom many in the resort town credited with the construction of the community's first permanent pier and school house – set out to become the next Reeve of Surrey Council (*Peace Arch News*, 12 July 2000; White Rock Museum and Archives 1993-30). His goal: to cancel the St Leonard Hotel's liquor license and shut it down for good. Cheering him on was the *Blaine Journal*, which felt that White Rock residents had become "dissatisfied" with the actions of the incumbent Reeve and the rest of Surrey Council in handling the St Leonard affair (14 November 1913).

With an election looming and residents of the municipality's growing resort town fearful of further diphtheria outbreaks, pressure on the incumbent Surrey Council to change its position vis-à-vis the St Leonard mounted. Just two weeks after the *Journal* announced Thrift's intention of running for reeve, John Keary, a long-time Surrey councillor and defender of the St Leonard, revealed to Blaine residents that the hotel's license would be cancelled in the coming weeks. According to the *Journal*, Keary admitted that "this course has been forced on the officials over there because of the agitation begun on [the US] side of the boundary ... The agitation has spread

... and has reached such tremendous proportions that the voice of the people must be heeded" (28 November 1913).

Even as it remained a rumour, word of the St Leonard Hotel's impending closure reverberated throughout Blaine. Estimating that three-quarters of city residents cared to see the bar shut down, *Journal* editor Sheets credited both Blaine and Surrey residents for seeing the fight through. "[Blaine residents] have hoped and prayed that this hell-hole be closed, but they were powerless to do anything," the *Journal* reported. "It is admitted that this announcement is due directly to the agitation that has started and has been continued from this side, later taken up by the good people over there" (28 November 1913).

In December, Surrey Council finally blocked the St Leonard Hotel's attempt to renew its liquor license. In an official announcement signed by the council's clerk at Cloverdale, BC, the license was set to expire on 15 January 1914. In celebration, the *Blaine Journal* listed everyone responsible for the closure, from Secretary of State William Jennings Bryan to Washington State senators Edward Brown and Miles Poindexter. Last but not least, Blaine's newspaper credited those Surrey residents responsible for the final and perhaps most important push to have the St Leonard's doors closed: "These gentlemen did their utmost to secure the cancellation of this license, and to blot out an international disgrace" (12 December 1913). Across the border in White Rock, news of the St Leonard's closure, which came to pass on the designated 15 January deadline, went unreported.

In the months and years that followed, the St Leonard Hotel issue, along with prohibition in general, continued to impact relations at the Blaine-White Rock border crossing. With the wind taken out of his sails by the incumbent Surrey Council's decision to cancel the St Leonard Hotel's liquor license, prominent White Rock resident Henry T. Thrift failed in his bid to defeat Reeve T.J. Sullivan in the spring 1914 election. Despite the setback, the *Semiahmoo Gazette* credited Thrift with being "productive of good results," and added that his campaign's "lasting benefit to the municipality will no doubt accrue from the many points raised by him at the meetings" (5 February 1914). In other words, Thrift's opposition to the St Leonard Hotel, although not always supported by the *Gazette*, had helped to re-establish amicable relations between Blaine and Surrey.

Figure 4.3 Although closed for business in 1914, the St Leonard Hotel remained next to the Peace Arch until it was demolished in the 1930s to make room for the Peace Arch Park (White Rock Museum and Archives, image 1995-13-183).

Before the end of 1914, Blaine's canneries ceased dumping salmon waste into Semiahmoo Bay (City of Surrey, 15 January 1915).[5]

MOVING BEYOND A PIONEER BOUNDARY

Still, there is evidence to suggest that residents of White Rock and Blaine, separated only by the narrow Semiahmoo Bay and two miles of open fields, continued to have dissimilar beliefs on the subject of alcohol. Almost two years after the St Leonard Hotel closed, reports that booze from Blaine was making its way into the Semiahmoo Indian reserve adjacent to White Rock threatened to open up old tensions between the two communities. On 15 November 1915, the *Semiahmoo Gazette* reported: "Hard cider from Blaine, Wash., is becoming a favorite beverage of the Semiahmoo Indian reserve," adding, "the good people of Blaine should look up the fyles [sic] of their 'Journal' regarding certain 'hell-hole' remarks." In response to anger from the Canadian side, the *Blaine Journal* noted that Surrey residents had every right to be upset. "They look upon it as much an injustice to them as the operations of the St. Leonard was to us. And they are right," the *Journal* admitted, calling on local authorities to be tougher on smugglers in the near and distant future (12 November 1915).

War clouds finally washed away British Columbia's booze in 1917, but White Rock remained one of only a few towns to oppose the bill. When Vancouver voted 8,412 to 5,696 in favour of provincial prohibition in September 1916, White Rock voters, few in number as they may have been, opted to remain wet by a vote of twenty-six to twenty-two (*Blaine Journal*, 22 September 1916). Among the cities and towns listed as participating in the vote, White Rock alone supported the liquor trade, while nearby Cloverdale, seat of Surrey Council, voted decisively in favour of prohibition.

How much of this resistance to prohibition in White Rock was fuelled by the St Leonard Hotel and salmon-dumping affairs is entirely debatable. What is evident, however, is that the issues revealed and contributed to ideas that this far-west border, drawn between US and Canadian pioneer communities, was real and mattered to the people who lived near it. For American social reformers in Blaine, the invisible international line delineated between two very different ideas of civilization. In the period after the St Leonard closed in 1914, Blaine Council passed several ordinances that affected minors, including a 8:00 p.m. curfew, a pool room ban, and a measure that disallowed them from entering second-hand stores or pawn shops (Blaine Municipal Government, 16 October 1916; 7 May 1917; 19 August 1918). The sale of fireworks for Fourth of July purposes was strictly prohibited in 1918 (Blaine Municipal Government, 18 February 1918). The council even passed ordinances to force local citizens to help police officers in need (Blaine

Municipal Government, 1 October 1917). In light of so much progressive legislation, it makes some sense that the community would have conceived of the border as an important barrier between itself and White Rock, which throughout the period prior to and during the First World War exploded in popularity as a resort destination. There, the White Rock Improvement League lobbied Surrey Council for better roads, beachfront toilets, and foot passageways over and under the Great Northern Railway line rather than focusing on social issues involving alcohol or the behaviour of local minors. Yet, the border was an important part of local affairs there as well, due in large part to Blaine's refusal to stop local canneries from dumping their waste into the shared Semiahmoo Bay. The St Leonard/salmon cannery standoff led to demands among some locals that the boundary running between Blaine and White Rock be more closely monitored. It was a request overwhelmed border agents could not fulfill, but it marked a substantial change in attitude for a region where Americans and Canadians had for generations passed effortlessly across a very permeable and above all peaceful boundary line (Fawcett 1913).[6]

NOTES

1 Evidence that Blaine residents frequented the beaches of White Rock can be seen in a report in 1912 of a Blaine boy drowning after wading into a hole at the mouth of the nearby Campbell River (Arbuckle et al. 1984, 67). The Roper family first settled on Semiahmoo Spit at Blaine in 1881 before patriarch Robert Roper applied for and received a land grant in White Rock, where he began a logging operation. When he died in 1889, the Ropers buried Robert in Blaine Cemetery.
2 Pioneer Surrey resident Margaret Stewart (n.d.) notes that during the 1880s many desperate Blaine residents crossed the line into Surrey to hunt for grouse and fish for trout, and virtually wiped them both out: "When the bottom fell out of Blaine then everybody was needing something," she states. "There was lots of people there and they come across. Well they just about cleaned [our] grous [sic] clean out. And they fished, they had nothing else to do for a while there."
3 A Surrey Licensing Board interview with Richard Asbeck reveals that in July 1911 the liquor licensing fee for the St Leonard increased from $200 per year to $400; *British Columbian*, "Surrey," 22 January 1918, 8. In his

address to Surrey Council, Reeve Sullivan noted that the municipality had erased its $8,000 deficit at the beginning of 1916 and had replaced it with a credit of $11,000, with most of the revenue coming from the sale of tax lands.

4 Luckily for travellers, the Canadian Railway Commission barred the GNR from doing so until 1925, when the rising ownership of automobiles made it a moot point.

5 In his annual report for 1914, Surrey Medical Health Officer Dr F.D. Sinclair announces: "The summer resorts of White Rock and Crescent [Beach] occasioned us less difficulty than formerly. The Provincial Government aided us here very materially by securing the abatement of the canneries nuisance on the shores of Semiahmoo Bay."

6 R. Fawcett to Malcolm R.J. Reid, 8 March 1913. Library and Archives Canada, MF C-10429 File 774004.

REFERENCES

Arbuckle, Marie, et al. 1984. *A Symbol of Our Heritage, The Old Fir Tree: Blaine Centennial History, 1884–1984*. Lynden: Profile Publications.

Berger, Carl. 1976. *The Writing of Canadian History: Aspects of English-Canadian Historical Writing: 1900–1970*. Toronto: Oxford University Press.

Blaine Municipal Government. City Council Minutes. NW369-6-1. Western Washington State Archives, Bellingham, WA.

City of Surrey. Surrey Council minutes. http://surrey.ihostez.com/Documents/DocumentList.aspx?ID=23297.

Ellenwood, Lorraine. 2004. *Years of Promise: White Rock, 1858–1958*. White Rock: White Rock Museum and Archives Society.

Fawcett, R. 1913. Letter to Malcolm R.J. Reid, 8 March. Ottawa: Library and Archives Canada (MF C-10429 file 774004).

Hill, Charles C. 1995. *The Group of Seven: Art for a Nation*. Toronto: McLelland and Stewart.

Immigration Department Correspondence. RG76 MF C-10429. Ottawa: Library and Archives Canada.

Moore, Stephen T. 2006. "Refugees from Volstead: Cross-Boundary Tourism in the Northwest during Prohibition." In *The Borderlands of the American and Canadian Wests: Essays on Regional History of the Forty-Ninth Parallel*, edited by Sterling Evans, 247–61. Lincoln: University of Nebraska Press.

Smith, Goldwin. 1891. *Canada and the Canadian Question.* Toronto: Hunter, Rose and Co.

Stewart, Margaret. Interview by Imbert Orchard, File 326: 1–2. Victoria: British Columbia Archives.

Stewart, Margaret. *How They Lived, as Told by Margaret M. Stewart.* Unpublished manuscript, 1959. Manuscript Collection, Surrey Archives, Surrey, BC.

White Rock Museum and Archives. "White Rock Elementary 70th Anniversary, May 19 & 20, 1984." W.R. Elementary School 70th Anniversary File, 1993–30.

5

International and Domestic Pressures on the Governance of the St Mary and Milk Rivers

MICHELLE MORRIS

Water, the common denominator of all living things, pays no heed to political boundaries. Tensions can ignite when neighbouring jurisdictions compete for limited resources. Such is the case for the province of Alberta and the state of Montana, which share the St Mary and Milk rivers. Both rivers arise in Montanan territory, but drain separately into Hudson Bay and the Gulf of Mexico. The St Mary flows north into Alberta where it eventually meets with the Oldman River and continues through Saskatchewan and Manitoba and into Hudson Bay. The Milk River, by contrast, meanders into southern Alberta before it returns to Montana and joins the Missouri River, which flows into the Mississippi River and into the Gulf of Mexico. A trans-watershed diversion, constructed within the Blackfeet reservation in the United States in 1917, hydrologically connects the separate river basins and ensures the flow of the Milk River for irrigation in Canada and the United States.[1] In dry years, 90 to 95 per cent of the Milk's waters originate in the St Mary watershed (Milk River Watershed Council 2008).

Downstream of the diversion in Alberta, the Milk River provides water for the municipalities of Coutts and Milk River in addition to the 400 commercial farms that operate within the watershed (Milk River Watershed Council Canada 2008, 39). In Montana, it supplies the communities of Havre, Chinook, and Harlem, as well as three reservations and approximately 800 farms (Daily and Cross 2004). The St Mary River, which diverts from Montana, supplies water to

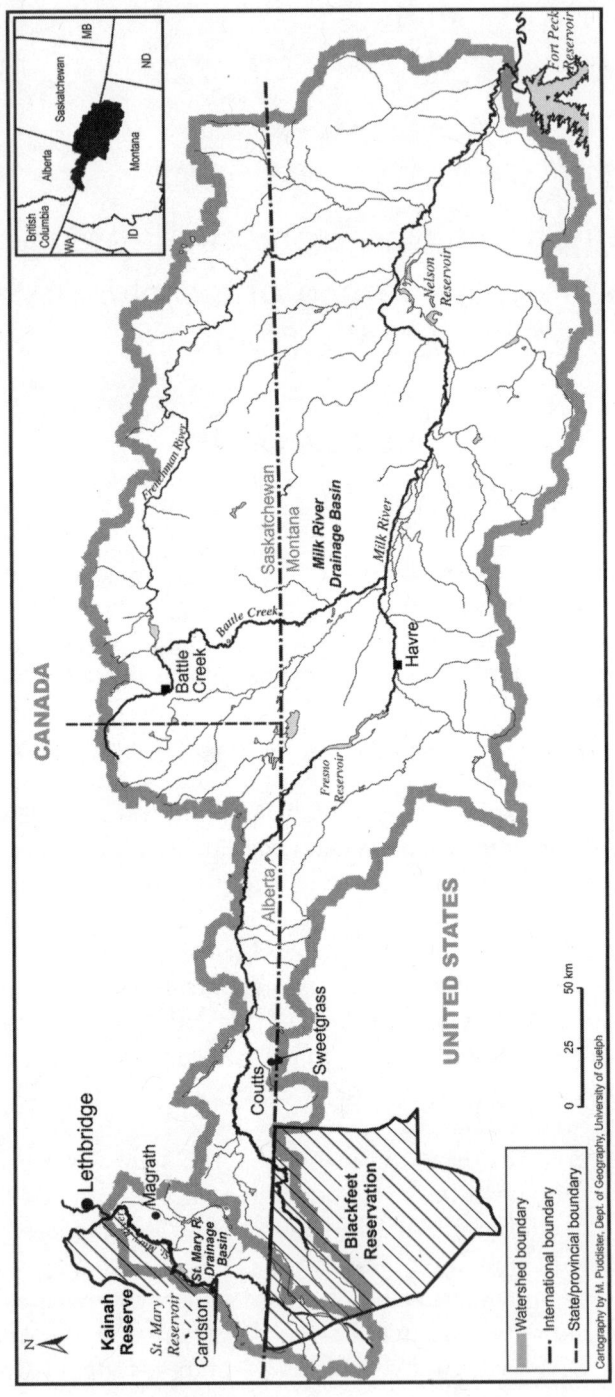

Figure 5.1 Map of the St Mary and Milk rivers (Puddister 2009, adapted with permission)

a number of municipalities, four irrigation districts, and the Kainah reserve within Canada (see figure 5.1 for a map of the watersheds).

The International Joint Commission (IJC) is an organization comprised of equal representation for Canada and the United States. The IJC deals with transboundary waters issues, and it created the 1921 Order on Apportionment (the order) for the shared St Mary and Milk rivers. The order dictates that during the irrigation season, when the flow of the St Mary is less than 666 cubic feet per second (cfs), Alberta has the right to 75 per cent of that flow. However, the order stipulates that when the flow is greater than 666 cfs, Alberta may take up to 500 cfs and any water in excess of 666 cfs should be divided equally between Alberta and Montana. The order allocates Montana the same proportion of water from the Milk River during the irrigation season. That is, Montana has the rights to three-quarters of the flow of the Milk River when its flow is less than 666 cfs and 500 cfs when the flow is greater than 666 cfs. The waters were to be divided equally during other times of year (International Joint Commission 1923). The St Mary River is snow-pack fed and has a much greater flow than the Milk, which depends on spring run-off and precipitation. The 1921 order gave Canada the greater flow from the two rivers – an arrangement Montana has never been satisfied with.

In 2003, Montana petitioned the IJC to reopen the order and determine whether or not each jurisdiction received its allocated volume of water. Governor Judy Martz (2003) requested that the IJC create a more equitable apportionment arrangement if it found that the concerned governments had not followed the agreement. The state claimed that it had not received its entitlement and noted increasing water shortages and new water agreements with the indigenous population as justification for reopening the order. Montana also questioned whether the then-eighty-two-year-old agreement was still practical. In response, Alberta opposed any alteration to the allocation scheme and maintained that the IJC should not revise it.

In this chapter, I offer a case study of this dispute by examining international and domestic factors related to the governance of these rivers. I will demonstrate that the IJC and the state of Montana have not impeded Alberta's ability to pursue its provincial self-interest related to the governance of the St Mary and Milk rivers, which is inspired by economic considerations. The IJC was never intended to serve as an international forum with expansive powers over domestic water governance. Furthermore, Montana unsuccessfully attempted to reopen the 1921 order. As international factors do not determine Alberta's water policies in regards to the St Mary and Milk rivers, it is necessary to turn to domestic variables to understand the province's governance of these rivers.

To disaggregate Alberta's domestic policy environment, I will focus on institutional, sectoral, and societal considerations. In terms of institutional actors, Alberta's government ministries play a key role in determining cross-border and domestic policy outcomes. Alberta Environment (AENV) and Alberta Agriculture and Rural Development (AARD) represent Alberta in bilateral Alberta-Montana forums, and policy makers from these ministries, which represent domestic interests, dominate discussions regarding the governance of the rivers. These institutional actors historically protected the interests of sectoral groups with water policies favourable to agricultural producers and processors. This support is evident in Alberta's protection of the 1921 Order on Apportionment, the subsidized rehabilitation of southern Alberta's expansive canal network and irrigation schemes, revisions to the Water Act to allow a water market to develop within the South Saskatchewan River Basin (SSRB), and a feasibility study regarding on and off stream storage on the Milk River. Southern Alberta is also home to various societal groups, such as environmental non-governmental organizations (ENGOs), that lobby for policy objectives although they are rarely recognized in policy outcomes. Additionally, two indigenous nations – including the Kainah, which borders the St Mary River in Canada, and the Blackfeet in Montana – have vital interests in the policy process. Sectoral prominence is illuminated when compared to the marginalization of ENGOs and indigenous interests. The determinants of Alberta's policy outcomes in regards to the St Mary and Milk River are government policy-makers who have historically acted in the economic interests of sectoral actors. Please see figure 5.2 for a summary of the argument I advance in this paper.

By disaggregating the domestic state, I offer an alternative perspective to unitary state-centred literature on transboundary water governance (Suhardiman and Giordano 2012; Dombrowsky 2007; Marty 2001; McCaffrey 1993). The following discussion of ENGOs and indigenous communities complements David LeMarquand's (1977) observations about the dominant role of the bureaucracy in Canada-US transboundary water governance. My argument reinforces Emma Norman and Karen Bakker's (2009) conclusions that

Actor	Degree of Influence	Support/Evidence
International – IJC	Partial	Involved in the process, but has not dictated unique outcomes in Alberta.
International – State of Montana	Partial	Unsuccessful in its attempt to reopen the 1921 Order on Apportionment.
Indigenous – Blackfeet Nation	Partial	Involved in the policy process. Represented on the current bilateral Alberta-Montana St Mary-Milk Rivers Water Management Initiative.
Indigenous – Kainah Nation	Minimal	Excluded from the bilateral Water Management Initiative. Claims of inherent water rights, aboriginal rights and title to water unrecognized within Alberta.
Sectoral – Agricultural Industry	Substantive	Inspired Alberta's protection of the 1921 Order on Apportionment, the subsidization of irrigation infrastructure, and water storage options for the Milk River.
Societal – ENGOS	Minimal	Unsuccessfully opposed the feasibility study for water storage in the Milk River Basin and lobbied for ecological objectives in the St Mary River.

Figure 5.2 Actor influence on Alberta

the devolution of authority to govern the St Mary and Milk rivers has not resulted in the empowerment of local actors. I also share Timothy Heinmiller's (2009) concerns regarding the vested interests that inhibit inclusive and flexible governance of the St Mary and Milk Rivers.

The evidence I discuss in this paper provides the case for concern around industry influence and the future ecological integrity of the Milk and St Mary River systems. Charles Lindblom (1982, 326) writes that policy makers are unlikely to regulate industries or reform institutions because to do so would induce a set of "automatic punishments" whereby fears of potentially slow economic growth and unemployment trump any benefits of institutional change or regulation. Alberta's stated reasons for opposing the order's re-examination were explicitly inspired by concerns of economic growth and stability, to the detriment of societal interests. Climate change impacts are expected to reduce stream flows in the St Mary River (AMEC 2009; Martz et al. 2007), which, in conjunction with water withdrawals for municipalities and agriculture, will further degrade the already threatened ecosystem. Climate change pressures suggest it is time to consider whether apportionment agreements made eight decades ago can meet the governance challenges of today. The extent to which the interests of the traditional owners and users of the Milk–St Mary River watersheds, the Blackfeet and Kainah, have influenced water governance in the transboundary context also suggests a need for more inclusive governance.

In terms of methodology in this study, I first engage the question of international intrusiveness. Specifically, to what extent is Alberta's water policy driven by external institutions, governments, or other actors? By answering this question it is possible to evaluate a number of relevant dependent variables, especially within Alberta. Institutional considerations include provincial ministries AENV and AARD. Industry associations and advisory groups, as well as their consultative links with provincial ministry officials, comprise sectoral actors. ENGOs are societal interests. The Kainah Nation in Alberta and Blackfeet in Montana are indigenous interests. The key to determining policy outcomes is to evaluate the autonomy of the state and the policy capacity of international factors without and sub-federal interests within the province of Alberta. To measure this policy impact, I will use a three-part scale of policy influence and state autonomy that ranges from minimal to partial to substantial.

My methods to find content for this analysis included reviewing historical sources, such as the 1915, 1917, 1920, and 1921 IJC hearings on the apportionment of the St Mary and Milk rivers. I reviewed contemporary government policy, along with the policy prescriptions of various sectoral and societal groups. I also examined the assortment of submissions that international, institutional, sectoral, and societal actors made to the IJC during its tour through Alberta and Montana.

This paper proceeds as follows. First, I discuss international factors including the IJC and the state of Montana. I will address the

Blackfeet and Kainah nations' influence, and then I will disaggregate the province of Alberta according to institutional, sectoral, and societal interests. The theme that threads these discussions together is Alberta's ability to maintain its autonomy by protecting sectoral interests, despite the pressures of international, societal, and indigenous actors.

THE BOUNDARY WATERS TREATY AND THE IJC

The 1909 Boundary Waters Treaty (BWT) is an international agreement between Canada and the United States that allowed the trans-watershed diversion. Article VI of the treaty deals explicitly with the St Mary and Milk rivers, and permitted the United States to divert water from the St Mary River into the Milk River, which flows through Canada before re-entering the United States. It also created a cooperative arrangement between Canadian and American bureaucrats to measure and apportion the waters. The IJC was created through this treaty, and is intended to oversee applications to alter the flow of transboundary waters. The IJC dictated the 1921 order. Importantly, the IJC has no enforcement mechanisms. It makes non-binding submissions to governments, which may or may not implement its recommendations. Article II of the BWT gives each party "the exclusive jurisdiction and control over ... all waters on its own side of the line which in their natural channels would flow across the boundary or into boundary waters," which prevents the IJC from interfering with domestic water policy.

THE IJC AND MONTANA'S REQUEST

The IJC responded to Montana's request to reopen the order by holding public hearings in July 2004 in Havre and Malta, Montana, Lethbridge, Alberta, and Eastend, Saskatchewan. Local media invited concerned individuals to attend and present their perspectives.[2] The IJC received approximately 115 submissions, which for the most part could be predicted along national lines: Americans wanted to reopen the 1921 order and Canadians did not (Faveri and Halliday 2007). Notable exceptions include environmental interests and submissions made by indigenous nations, which I will explore later in this chapter. In response to the St Mary-Milk submissions and hearings, the IJC directed that a task force, comprised of civil servants from both sides of the border and an engineer from a private consulting firm, analyze the apportionment and offer advice for potential improvements (Bourget and Claman 2004).[3]

The task force produced a report of final recommendations in April 2006. The report focussed on technical issues, including methods of measurement and administrative measures such as accounting for surplus and deficit deliveries of water. The report states that from 1950 to 2004, Alberta received 55 per cent of the combined flow of the St Mary and Milk rivers and several other smaller tributaries, whereas Montana received 45 per cent (International Joint Commission 2006a). Due to considerable government and private investment, Alberta had a much more water-efficient irrigation system than Montana. Montana lacked the infrastructural capacity required to divert its allocated volume from the St Mary, and Alberta was unable to divert its allocated flow from the Milk for the same reason. The task force failed to reach consensus on any key recommendations, which indicates the value both jurisdictions placed on achieving their own policy objectives.

In 2008, Montana and Alberta policy makers reopened discussions around the St Mary and Milk rivers through the Montana-Alberta St Mary and Milk Rivers Water Management Initiative. The revival of discussions was in part a response to an IJC request that consultations reconvene, and in part a mutual desire to cooperate. The IJC has a limited role in these discussions, as no IJC representatives are present in bilateral meetings. Priorities for the new group include rehabilitating the diversion infrastructure within Montana and working together on balancing periods, the units of time by which water flows are measured, through more frequent cross-order consultations (Montana and Alberta 2008). It also has a mandate to discuss irrigation and in-stream flow needs (Montana and Alberta 2008). This group's meetings in 2009 and 2010 indicate that solutions will likely involve improving infrastructure on both sides of the border to allow both Montana and Alberta to receive their full allocations from the rivers (Montana and Alberta 2009; 2010).

US PRESSURES

Montana has the ability to influence Alberta's policy on the St Mary and Milk rivers. Geographical location means the state will necessarily be considered in the policy process. However, Alberta has successfully resisted attempts to reopen the order, despite Montana's preference to the contrary. The state has not dictated unique policy outcomes within Alberta, but it is involved in the process behind them.

The most pressing issue for Montana is the infrastructure that diverts water from the St Mary to the Milk River, which is in dire need of repair. Its failure would adversely affect the Blackfeet Nation and the populations that reside within the Milk River watershed on both sides of the border. In the beginning of the spring 2008 irrigation season, a section in the 47 kilometres of steel siphon that diverts water into the Milk River leaked because of expansion and contraction of the pipe (Steiner 2008, 4). The St Mary Working Group, a group of ranchers and politicians, has unsuccessfully lobbied Congress for funding to fix the diversion.

If Montana were to rehabilitate the broken section of siphon, there could be several implications for Alberta's policy environment. First, increasing water diverted from the St Mary River watershed into the Milk River could threaten the already degraded condition of the St Mary River, which is 118 per cent allocated within Alberta (Alberta 2005). However, the diversion could potentially provide water for residents within the Milk River basin who seek a supplemental supply. In meetings, the Alberta-Montana Water Management Initiative has identified rehabilitating the diversion as one of its main consideration, and a potential point at which the dispute could be resolved (Montana and Alberta 2010).

It is necessary to discuss international considerations because they form part of the policy process that produces policy outcomes. However, the IJC and the state of Montana have not determined unique policy outcomes within Alberta. The IJC was never intended to intensively pressure domestic governments to produce its own policy outcomes. It has not made any recommendations to Montana and Alberta regarding the St Mary and Milk rivers. Instead, it has generated data and encouraged policy-makers within the state and provincial governments to resolve the contentious issue. The state of Montana was unable to reopen the 1921 order and revise the apportionment arrangement.

INDIGENOUS INTERESTS: THE BLACKFEET AND KAINAH NATIONS

Indigenous tribes first inhabited the St Mary and Milk River watersheds at least 8,000 years ago (Glenn 1999, 151). The watersheds are part of the traditional territory of the Blackfoot Confederacy, which includes the Blackfeet Nation in Montana, as well as the Kainah, Siksika, and Pikiini nations in southern Alberta. In a submission to the IJC in 2004, the Kainah Nation pointed out that "the Canada-US border was surveyed ... without regard to the Traditional Territory or the Blackfoot Confederacy's social, political, and economic relations" (Shade 2004). Likewise, the BWT and the 1921 order failed to consider the rights of the current residents of the watersheds, the Blackfeet and Kainah nations. Both have voiced dissatisfaction to the IJC because of this failure (Whiteing 2004; Shade 2004).

The Blackfeet Nation in Montana has reserved rights to water as articulated in the 1908 *Winters v. United States* decision of the United States Supreme Court. The Winters doctrine recognizes an implied water right in the creation of reservations, and required that the US federal government ensure water needs of reservations were met (*Winters v. United States*, 207 US 564, 1908). The headwaters of both the St Mary and Milk rivers and the diversion scheme are within Blackfeet reservation territory, while the diversion scheme carries water out of it. The tribe has been negotiating its water rights with the state of Montana for over twenty-five years. In spring 2009, the tribe and the state agreed on the Blackfeet Water Compact, which assigns $4 million to repair and enlarge a reservoir on the reservation and allocates 50,000 acre-feet of the St Mary River to the Blackfeet Nation. Montana legislature unanimously passed the Blackfeet Water Compact Bill in March 2009, and it now requires the approval of the US Congress before receiving the tribe's final consent. At the time of writing this chapter, the bill has been in a Congressional committee for over a year. The 1921 order was not a part of tribal-state negotiations and Alberta's allocation will not be affected by the final outcome of the Blackfeet Water Compact. However, the allocation to the Blackfeet could impact water allocations to ranchers in the St Mary and Milk River watersheds within the United States, as allocations to farmers and ranchers in Montana were made before Indian reserved rights were considered.

The Blackfeet Nation has a representative on the Alberta-Montana St Mary-Milk Rivers Water Management Initiative. Any resolution that does not consider the Blackfeet's reserved rights will be unacceptable to tribal leadership, which states "the very validity of the Boundary Waters Treaty is questionable in light of its failure to consider Indian reserved rights" (Whiteing 2004). In fact, the Blackfeet's reserved water rights are not a topic of discussion for members of the

Water Management Initiative. Members of the nation are involved in the policy process, which shows it has some influence. It has not forced Alberta to adopt unique policies, but it has pressured the state of Montana to reconsider its water management and the 1921 order. Though the Blackfeet successfully negotiated with Montana for water rights, US Congress has yet to approve of the agreement. It is possible that its cost may prevent timely congressional approval.

Members of the Kainah Nation, which borders the St Mary River in Canada, claim riparian rights and aboriginal rights and title to the St Mary waters (Phare 2009). Section 35(1) of the Canadian constitution protects aboriginal title, and Canadian case law has produced a number of related precedents. In Canadian jurisprudence, aboriginal title is considered a right to land, and indigenous peoples traditionally view this right as coupled with stewardship responsibilities. The authority to govern water according to indigenous laws and customs, rather than ownership per se, is the issue. Aboriginal rights, on the other hand, consist of culturally integral site-specific practices. In the case of water, this may include rights to fish, navigate, or use water for spiritual purposes (Passelac-Ross and Smith 2010). This right requires water of an adequate quantity and quality to fulfill these cultural purposes. The fact that the Kainah Nation claims these rights is significant. If its argument is proven and accepted, the result could considerably alter the way that indigenous peoples in southern Alberta are involved in water governance. In addition to these rights, the Kainah Nation claims it never ceded rights to water through the Canadian treaty-making process. Treaty 7, which created the reserve, contains no mention of water.

The Kainah Nation is also involved in irrigated agriculture through the Blood Tribe Agricultural Project (BTAP) that was created with the federal and provincial governments (Shade 2004). The Kainah were assigned water to irrigate up to 25,000 acres of land and $7 million to promote agriculture on the reserve. Irrigation water comes from the St Mary-Belly-Waterton diversion within Alberta. Claims of aboriginal rights and title, contested treaty rights to water, and BTAP mean that the Kainah are an important actor in southern Alberta's water governance. The Kainah Nation does not have a representative on the Water Management Initiative that is considering resolutions to the St Mary and Milk rivers dispute. The Kainah Nation is affected by the governance of the St Mary and Milk rivers but has not been involved in producing policy, and its claims of inherent water rights have not been recognized. This suggests marginalization from the policy process, and that the Kainah Nation is minimally influential on a substantively autonomous state. However, the unresolved claims of aboriginal rights and title to water may give the tribe a significant role in the future.

Indigenous nations have vital interests in the way that the St Mary and Milk rivers are governed and in any potential resolution that will come from the bilateral group. Despite these interests, they have uneven representation within bilateral forums. As international considerations and indigenous interests are not determining provincial water policy, it is necessary to turn to Alberta – and the institutional, sectoral, and societal interests within it – to understand its water policy outcomes regarding the St Mary and Milk rivers.

INSTITUTIONAL FACTORS

The Provincial Bureaucracy and AENV

Within Alberta, water governance falls under the ambit of several bureaucratic ministries, including AENV and AARD. AENV is most directly involved in St Mary and Milk River water issues. The Southern Region Water Management Operations branch of AENV owns and operates the main diversion infrastructure from the St Mary River. This system delivers water from the Rocky Mountains eastward and provides water for four irrigation districts. AENV also administers water licenses under Alberta's Water Act.

In the mid-1990s, the provincial government revised the Water Act to establish a legislative framework that would allow water markets to develop in Alberta's river basins. Currently, the only river basin in which water licenses have been sold is the South Saskatchewan River Basin (SSRB), home of the St Mary River. AENV policy makers were heavily involved in the revisions, and play key roles in the approving license transfers. A senior AENV civil servant has the authority to accept or reject water license trades and allocation transfers and hold back up to 10 per cent of transferred waters to meet ecological objectives. These features of the Water Act lend themselves to significant bureaucratic interpretation, which reinforces AENV's dominant role governing the St Mary and Milk rivers. The AENV's southern region branch was responsible for the 2002 decision to close the St Mary, Belly, and Waterton rivers to any further allocations and the 2006 decision to close the entire SSRB to any further licensed allocations (Alberta 2003; 2006). This closure indicates the ecological stress new diversions would cause, and also made license transfers requisite. AENV closed the Milk River to any further irrigation and industrial licenses in 1985. The environment ministry also commissioned a feasibility study for on and off stream storage options on the Milk (Anderson et al. 2009).

Given the broad authority AENV holds as owner of infrastructure and deliverer of water allocations, it is unsurprising that it has been directly involved in IJC facilitated discussions and bilateral Alberta-Montana initiatives. After it heard of Montana's request to have the 1921 order reopened, the ministry consistently opposed any alteration to the apportionment formula. Environment Minister Lorne Taylor (2004) articulated the perspective maintained by environment ministry employees as he wrote to IJC commissioners: "Stability and certainty are needed for investment and economic growth, and even the perception that the Order might change threaten that stability." Taylor's statement indicates the economic concerns that inspired maintaining the order and that Alberta was unwilling to concede Montana's request. AENV voiced these concerns at IJC held stakeholder consultations and within the task force.

AARD

AARD also has an important role in Alberta's water policy. The ministry delivers public funds to rehabilitate irrigation district water conveyance infrastructure through the Irrigation Rehabilitation Program. In 1969, the Alberta government adopted this cost-sharing agreement with the districts to encourage greater efficiency and enhance water conservation. The current arrangement has the province funding 75 per cent of infrastructure rehabilitation and the districts the remaining 25 per cent of costs. The Alberta government and irrigation districts have contributed approximately $655 million to irrigation district rehabilitation (Alberta 2004). The St Mary River Irrigation District (SMRID), the largest district that uses St Mary River water, received an average of $6.2 million annually between 2006 and 2010 to rehabilitate irrigation infrastructure through the Irrigation Rehabilitation Program (Alberta 2012).

Four of Alberta's thirteen irrigation districts convey St Mary River water to users, including the Magrath Irrigation District (MID), the Raymond Irrigation District (RID), the SMRID, and the Taber Irrigation District (TID). The SMRID delivers water to approximately

1,900 water users, ten municipalities, and fifteen Hutterite colonies. It conveys water using a 2,000 km network of canals and pipelines and provides water to irrigate approximately 372,000 acres of land (Alberta 2004, 8).

As with AENV, AARD presented to the IJC as it toured through southern Alberta. It also has representation on the current bilateral Alberta-Montana group on water management. AARD's stance paralleled AENV in that it urged for the continued observance of the 1921 order. Policies designed and implemented by AARD conform to the interests of irrigators and the value-added processors that reside within the St Mary and Milk River watersheds.

SECTORAL CONSIDERATIONS

Sectoral interests within the St Mary and Milk River watersheds are predominantly agricultural. Diversions from the St Mary provide water for numerous crops such as sugar beets, potatoes, corn, peas, and beans. These crops are sold to value-added industries and processed, packaged, and exported to domestic and international markets. Water from the river is also used to grow forage for livestock intended for domestic consumption and export. Economists Kurt Klein and Lorraine Nicol (2006, 93) identify the forward-backward links between equipment suppliers, producers, and processors as an important feature of the economy within the SSRB. Agricultural uses of St Mary water patterns the increasing significance of the value-added industry and the consolidation of food production into fewer, larger centres, which characterizes Canada's agricultural production (Skogstad 2008, 503). As value-added industries employ more people than the primary producers that grow the products, they have become increasingly important to policy-makers concerned with economic productivity. AARD points out that "30 percent of regional employment opportunities in southern Alberta are directly or indirectly associated with irrigated agriculture" (Alberta 2004, 2). These considerations led groups such as the Canadian Federation of Agriculture and Alberta Sugar Beet Growers to lobby the IJC not to reopen the 1921 order and also inspired AENV and AARD's stance (Friesen 2004; Harris 2004; Oudman 2004).

Value-added industries that use water to process commodities grown with St Mary water include Canbra Foods, Spitz Seeds, Sakai Spice, Lucerne Foods, McCain Foods, Hostess-Frito Lay, Lamb-Weston

Processors, Sunrise Poultry, Maple Leaf Hog Processors, Maple Leaf Foods, and Rogers Sugar. Taber's Rogers Sugar Refinery, the only plant of its kind in Canada, is completely dependent on irrigation-delivered water for the sugar beets it processes (Korevaar 2004). Two potato-processing plants opened in southern Alberta after the most recent revisions to the Water Act. In their analysis of the 2001 water market, Nicol and Klein (2006, 95) note the significance of the transfer of allocations from low- to high-value crops, such as from forage to sugar beets or potatoes. The high-value crops are processed, which requires large quantities of water. Crop producers, then, want to ensure processors have access to water so that they can sell their crops. The policies designed by the AENV policy-makers exhibit substantive sectoral influence. By protecting the 1921 order, Alberta's bureaucrats have acted in the interests of the sectoral actors that depend on the international water apportionment arrangement.

In the Milk River Basin, water is used to grow cereal and forage crops and to sustain ranches. One important difference between the St Mary and Milk River watersheds is the lack of processing plants within the latter, which means that ranchers and irrigators are the dominant sectoral actors. According to the Milk River Watershed Council Canada, 59 per cent of employment in Alberta's portion of the St Mary watershed is based in agriculture, and 40 per cent of that agriculture is related to cattle ranching (Milk River Watershed Council Canada 2008, 36, 41). Irrigation accounts for 93 per cent of licensed water within from the Milk River (60). Pressure from ranching and irrigation interests within the basin led AENV to commission a feasibility study for on and off stream water storage. This indicates substantive sectoral pressure on the state, and a noteworthy role for Milk River ranchers in Alberta's policy process regarding the St Mary and Milk rivers. A review of a Milk River Basin Water Management Committee document is telling. The committee suggests "a storage site on the Milk River ... provides the largest range of benefits to basin residents in Alberta" (Gilchrist 2004, 1).

SOCIETAL CONSIDERATIONS

The Southern Alberta Group for the Environment (SAGE) and the Alberta Wilderness Association, two Albertan ENGOs, have both voiced strong policy positions on the St Mary and Milk rivers. SAGE points out "the aquatic environments of both the lower St Mary River in Alberta and the lower Milk River in Montana are under stress and considered to be degrading due to current water management" (Jericho 2004). When AENV commissioned its on- and off-stream storage feasibility study in 2003, SAGE and AWA opposed it. AWA wrote that a study on an issue that would "provide water for low value irrigation projects and damage a high value conservation area at a cost of hundreds of millions of dollars to the Alberta taxpayer" should not proceed (Bray 2003, 4). SAGE and AWA had minimal influence on the province of Alberta, as their concerns were not a factor in the decision to conduct a study for water storage.

Both AWA and SAGE urged the Alberta government to consider in-stream flow needs and preserving the ecological integrity of the rivers. However, the St Mary River is currently 118 per cent allocated, which suggests they have had a minimal impact on policy. The marginalization of environmental interests stands in stark contrast to policies that favour industry, including the development of a water market, the subsidization of irrigation, and conducting a feasibility study on water storage on the Milk River.

In conclusion, international factors such as the IJC and the state of Montana do not determine Alberta's water policy on the St Mary and Milk rivers. The IJC lacks enforcement mechanisms, and Montana has been unable to make changes to the 1921 water allocation agreement. These international factors, then, were merely a part of the policy process and only partially influenced Alberta. Provincial policy makers, from AENV and AARD, determine policy outcomes for the St Mary and Milk rivers. Policy outcomes have historically been a result of complementary interests of sectoral actors within the watersheds. By amending Alberta's Water Act to allow a water market to develop within the SSRB, which favours value-added industry and high efficiency irrigation equipment, AENV aligned with the goals of producers and processors of high-value specialty crops. This ministry has also acted to promote the goals of industrial actors in the Milk River Basin by conducting feasibility studies for on- and off-stream storage. Montana's weakness in this dispute was its dilapidated irrigation infrastructure; the subsidization apparent in Alberta is unparalleled in Montana. The fact that the dispute is heading towards a resolution that involves improving infrastructure on both sides of the border, to make more water available for each party, benefits the sectoral interests that reside in the borderlands.

NOTES

1 This diversion scheme includes 47 kilometres of canal, two sets of steel siphons, and five concrete drop structures, which divert water from the Sherburne Dam and St Mary Diversion Dam within Montana into the north fork of the Milk River and into Canada.
2 The IJC published these consultations on its website (www.ijc.org/rel/boards/smmr/public_consult-e.htm). Unless noted otherwise, citations described as submissions to the IJC are available at the website.
3 Several Native American and First Nations groups, including the Chippewa-Cree, Assiniboine-Gros Ventre, and Kainah Nations, were invited to attend the meetings as observers.

REFERENCES

Alberta. 2003. Alberta Environment. *Apportionment of the St Mary and Milk River Flows Under the Boundary Waters Treaty: Background Information on Alberta's Position.* 14 August.
– 2004. Alberta Agriculture, Food, and Rural Development. *Irrigation Development in Alberta: Water Use and Impact on Regional Development, St Mary River and "Southern Tributaries" Watersheds.* August.
– 2005. Alberta Environment, Southern Region. *South Saskatchewan River Basin Water Allocations: Background Study for the South Saskatchewan River Basin Plan.*
– 2006. Alberta Environment. *Approved Water Management Plan for the South Saskatchewan River Basin.* August.
– 2012. Alberta Agriculture and Rural Development. *Irrigation Rehabilitation Program Status Reports.*
AMEC. 2009. *South Saskatchewan River Basin in Alberta: Water Supply Study.* Lethbridge, AB: Alberta Agriculture and Rural Development.
Anderson, Marv, Klohn Crippen Berger Ltd, and Milk River Watershed Council Canada. 2009. Milk River Supplemental Water Supply Investigation. *Research and Science Note 3.*
Bourget, Elizabeth, and Murray Claman. 2004. *Directive to the International St Mary-Milk Rivers Administrative Measures Task Force.* 30 November. http://www.ijc.org/conseil_board/st_mary_milk_rivers2/en/smmr2_mandate_mandat.htm.
Bray, Shirley. 2003. "Keep the Milk River Wild – No Dams!" *Wild Lands Advocate* 11 (1): 4–6.

Daily, Mike, and Mary Cross. 2004. "Water Supply and Distribution." In *The Milk River: International Lifeline of the Hi-Line, A Guidebook*, 9–11. Montana: Montana Department of Natural Resources and Conservation, Bureau of Reclamation, Milk River International Alliance (Milk River International Alliance, IJC submission, July).

Dombrowsky, Ines. 2007. *Conflict, Cooperation and Institutions in International Water Management: An Economic Analysis*. Northampton: Edward Elgar.

Faveri, G., and R. Halliday. 2007. "The St Mary and Milk Rivers: The 1921 Order Revisited." *Canadian Water Resources Journal* 32 (1): 75–92.

Friesen, Bob. 2004. Canadian Federation of Agriculture, Submission to the IJC, 25 August.

Gilchrist, Tom. 2004. Milk River Basin Water Management Committee Submission to the IJC. 29 July.

Glenn, Jack. 1999. *Once upon an Oldman: Special Interest Politics and the Oldman River Dam*. Vancouver: UBC Press.

Harris, Merill. 2004. Alberta Sugar Beet Growers submission to the IJC. 9 August.

Heinmiller, T.B. 2009. "Managing Water Scarcity in the Prairie Region: The Role of the IJC in a Changing Climate." In *Transboundary Environment Governance in Canada and the United States*, 19–34. Woodrow Wilson Canada Institute Occasional Paper #3.

International Joint Commission. 1909. *Treaty Between the United States and Great Britain Relating to Boundary Waters, and Questions Arising Between the United States and Canada*.

– 1915. *Hearing and Argument in the Matter of the Measurement and Apportionment of the St Mary and Milk Rivers and Their Tributaries held in St Paul, Minnesota in May 1915*. Washington: Government Printing Office.

– 1923. *In the Matter of the Measurement and Apportionment of the Waters of the St Mary and Milk Rivers and Their Tributaries in the United States and Canada*. Washington: Government Printing Office.

– 2006a. *International St Mary-Milk Rivers Administrative Measures Task Force: Report to the IJC*. http://www.ijc.org/rel/pdf/SMMRAM.pdf.

– 2006b. *Overview of Public Review and Comment Process for the April 2006 Task Force Report*. http://www.ijc.org/rel/boards/smmr2/2006report/overview.htm.

Irrigation Water Management Study Committee. 2002. *South Saskatchewan River Basin: Irrigation in the 21st Century*. Lethbridge: Alberta Irrigation Projects Association.

Jericho, Klaus. 2004. Southern Alberta Group for the Environment Submission to the IJC. 20 July.

Korevaar, Arie. 2004. Alberta Sugar Beet Growers Submission to the IJC. 24 July.

LeMarquand, David. 1977. *International Rivers: The Politics of Cooperation*. Vancouver: Westwater Research Centre, UBC Press.

Lindblom, Charles. 1982. "The Market as Prison." *The Journal of Politics* 44: 324–36.

Lystrom, David, and R.A. Halliday. 1995. *Report to the International Joint Commission on the Division of the Waters of the St Mary and Milk Rivers*. Helena: US Geological Survey.

Marty, Frank. 2001. *Managing International Rivers: Problems, Politics, and Institutions*. New York: Peter Lang.

Martz, Lawrence, Joel Bruneau, and Terry J. Rolfe. 2007. *Climate Change and Water – SSRB Final Technical Report*. Ottawa: Climate Change Impact and Adaptation Program, Natural Resources Canada.

Martz, Judy. 2003. Letter to Dennis Schornack, Chairman, United States Section of the International Joint Commission. 10 April. http://www.ijc.org/rel/boards/smmr/public_consult-e.htm.

McCaffrey, Stephen. 1993. "Water, Politics, and International Law." In *Water in Crisis: A Guide to the World's Freshwater Resources*, edited by Peter Gleick, 92–100. New York: Oxford University Press.

Milk River Watershed Council Canada. 2008. *State of the Watershed, 2008*. Milk River: Milk River Watershed Council.

Montana. 2008. Department of Natural Resources and Conservation. *The State of Montana's Comments and Recommendations on the International St Mary-Milk Rivers Administrative Measures Task Force Report*. June.

Montana and Alberta. 2008. *Montana-Alberta St Mary and Milk Rivers Water Management Initiative*. Terms of Reference. November.

– 2009. *St Mary & Milk Rivers Water Management Initiative Joint Initiative Team Meeting #10*. Meeting Minutes. December.

– 2010. *St. Mary & Milk Rivers Water Management Initiative Joint Initiative Team Meeting #12*. Meeting Minutes. February.

Nicol, Lorraine, and K.K. Klein. 2006. "Water Market Characteristics: Results from a Survey of Southern Alberta Irrigators." *Canadian Water Resources Journal* 31 (2): 91–104.

Norman, Emma, and Karen Bakker. 2009. "Transgressing Scales: Water Governance across the Canada-US Borderland." *Annals of the Association of American Geographers* 99 (1): 99–117.

Oudman, Rob. 2004. *Alberta Vegetable Growers (Processing) Submission to the IJC.* 20 August.

Passelac-Ross, Monique, and Christina M. Smith. 2010. "Defining Aboriginal Rights to Water in Alberta: Do They Still 'Exist'? How Extensive are They?" *Canadian Institute of Resource Law Occasional Paper* #29. April.

Phare, Merrel-Ann S. 2009. *Denying the Source: The Crisis of First Nations Water Rights.* Calgary: Rocky Mountain Books.

Puddister, M. 2009. "Figure 6: St Mary and Milk River Basins." In R.C. de Loë, *Sharing the Waters of the Red River Basin: A Review of Options for Transboundary Water Governance,* 45. Guelph: Rob de Loë Consulting Services.

Shade, Chris. 2004. Kainah/Blood Submission to the IJC. 29 July.

Simonds, Joe. 1999. *The Milk River Project.* Denver: Bureau of Reclamation History Program. http://www.usbr.gov/dataweb/html/milkrive.html.

Skogstad, Grace. 2008. "Canadian Agricultural Programs and Paradigms: The International Influence of Trade Agreements and Domestic Factors." *Canadian Journal of Agricultural Economics* 56 (4): 493–508.

Steiner, Al. 2008. "St Mary Canal Start up a Challenge." *Milk River Watershed News.* Spring.

Suhardiman, D. and M. Giordano. 2012. Process-Focused Analysis in Transboundary Water Governance Research." *International Environmental Agreements* 12 (3): 299–308.

Taylor, Lorne. 2004. Alberta Environment Minister to Herb Gray, IJC Chair, Canadian Section and Dennis Schornack, IJC Chair, American Section, 13 July.

Whiteing, Jeanne (Blackfeet Nation). 2004. Submission to the IJC, 7 September.

6

Water and Political Relations between the Upper Plains States and the Prairie Provinces: What Works, What Doesn't, and What's All Wet

PAUL R. SANDO

The Red River basin and the attendant basins of the Souris-Assiniboine Rivers straddle the Canada-US border and create a number of issues that require substantive cooperation and relations between the affected states and provinces. In this paper, I focus on three areas of contention and cooperation: the Souris River Basin flood control project and Northwest Area Water Supply Project, the Devils Lake Drainage Project, and the Red River Basin flood control efforts. These projects deal with attempts to control water in excess, and one, the Souris basin and the NAWS project, deals with water resources that move between basins. All three also highlight different aspects of relations relative to proximate and physical geography and are examples of what works, what doesn't, and what is a work in progress (or, what's all wet).

The cross-border negotiation and argument over what to do and what is possible with water resources is paramount. Relations between states, provinces, countries, and tribes are all at stake. Many of the difficulties in geography also present difficulties in relations. All possibilities and discussions hinge on what to do – and indeed what is humanly, economically, and physically possible to do – with excess water. Who has the right to do what with water resources, and how should they mitigate the effects on others or not? What are the implications of, limitations of, and issues with geography and the physical environment? How does the human geography of the region and of North America affect the problem? Those in the region can usually understand most of the situation, but what happens when those in Ottawa, Washington, and elsewhere get involved?

These projects and the areas they affect have seen some momentous and monstrous floods in what has been stated to be the wettest succession of seasons in recorded memory. The Red River saw major flooding in 2009, 2010, and 2011. The Souris River basin experienced a major flood in 2011 for the first time since 1969. Devils Lake continues to increase in size as it floods out its basin, a process that has taken many years to come to a head. All of these areas share a similar problem: water in excess. But they also share their engineering for arid seasons and water scarcity. Each project has many attendant environmental concerns; significant human activity and presence; and similar geographic characteristics in the flat, glacially modified terrain of the northern Great Plains. None of these factors has helped the complexity or the enormity of the problems in the area.

COMPLEXITY, ENVIRONMENTAL CONCERNS, AND GEOGRAPHY

Basins and water projects that straddle divides and international borders create a number of issues that require substantive cooperation between states and provinces. Three major issues create conflict around the movement of water in the affected region: efforts to control and mitigate excess water, environmental protectionism at the expense of other regions, and local alterable and inalterable geography. If the three mitigation projects can align their positions on these topics, each may eventually reach an outcome that protects both the human geography and also the environment.

Project complexity also affects success. The simpler the goals of a project, the more successful it is, as evidenced by the Souris Basin and the international board that governs it. As one result of this, other entities have begun to take a step-by-step approach. For example, a project might only address flood control and leave other concerns for later. This type of simplicity of design is, fortunately or unfortunately, increasingly less possible. Still, less complex projects have proven to be more successful than more comprehensive ones.

One of the main characteristics of these water projects, and a main source of complexity, is concern over the environment. More groups that represent more diverse interests and regional impacts are voicing more environmental concerns than ever before. In the past, the good of the region often came at the expense of individual interests. Recent political and economic changes have tried to make sure that projects take all interests into account and lead to the greatest benefits with the least amount of risk to most parties.

The nature of the regional geography is behind all of these projects and any controversy or cooperation that goes along with them. Any action will always be judged as success or failure based on what the land and environment will accept. Mother nature can sorely test even the most massive projects. In 2011, an unusually heavy winter snow and wet spring tested the Missouri River and its systems of huge dams for flood control, water retention, and power generation. The result was massive flooding in that basin that persisted into 2011 despite the best human efforts. The characteristics of the region are the first topic of discussion.

WATER AND THE NEIGHBOURHOOD: IT IS REALLY, REALLY FLAT

The geography of the Red River basin ensures that the problems of one neighbour – be it a state, province, or country – is another's business. In this region, water plays the central role as the object, cause, effect, and result of political and economic conflict, usually over where excess water goes, that involves several cultural groups, at least two states, at least three provinces, and two nations. An engineer involved in the Souris Basin project reveals the misunderstanding that curses all four areas: "It is easier to engineer for too much water, than it is for too little" (International Joint Commission 2002). Most of the people involved in the Souris River Basin flood control project and Northwest Area Water Supply Project, the Devils Lake Drainage Project, and the Red River Basin flood control efforts would likely beg to differ.

The Red River basin itself is the result of ice sheets that were present during the last round of continental glaciation in North America. The ice sheets not only altered the paleodrainage of the region, but also ground and graded the much of the land down to a gently rolling terrain. After the glaciers retreated, several large lakes occupied part of the region from about 7,000 to 5,000 BCE, and their beds contributed to the flatness. The glaciers left behind moraines and other deposits that disrupted local drainage and caused problems later on. The region also has a unique, and major, geographic issue: all of its water resources except for the Missouri system drain northward and eventually into Hudson Bay. Due to the local climate, northward drainage poses a whole host of problems related to flood control and flow.

The geographic constraints of the region and its terrain limit what can be done with any excess water. If it is removed from one place, there are ramifications in another that are often felt many miles distant. In the case of the Red River flood control efforts, one official put it thus: "Imagine a large conference table as your valley. Now gouge a half inch deep, crooked, meandering channel. Then elevate one end of the table one inch. Now slowly pour a bucket of water in that channel. The water will get to the low end of the table, but it will take a long time to do so and affect most of the area" (Conan 2010). Additionally, the river and bay into which the water drains are frozen, which has created serious back-up problems.

Human geography also plays an important role in the discussion of these three projects. People have different expectations and methodologies for dealing with geographic constraints. The Canadian government simply operates differently than the US government. The states are different from each other and from the provinces. The autonomy of tribal interests makes addressing their concerns akin to managing relations between sovereign nations. There are also rural and urban issues to address, in addition to environmental concerns such as changing water quality and possible cross-basin contamination.

When people or interests from outside the region begin to take an interest in these issues, they introduce a whole new level of complexity and conflict. Though regional parties may understand the problem, the region, and the parties involved and propose a potential solution, outsiders often do not. For example, in a finding in regards to one of North Dakota's actions on a project, a US federal judge incorrectly referred to a "mountain range" (United States 2009) despite the fact that there are no mountains in North Dakota except by name, and the main area of elevated terrain is in the southwest corner of the state, well away from the drainage basins under discussion.

Cross-border negotiations and arguments over what to do and what is possible with water resources are paramount. Relations between states, provinces, countries, and tribes are all at stake, and many of the difficulties in geography also strain these relationships. All possibilities and discussions hinge on what is done with excess water, and indeed what is humanly, economically, environmentally, and physically possible to do with it. Who has the right to do what with water resources, and how responsible must they be for mitigating the effects on others?

THE SOURIS RIVER BASIN

The Souris River is a small meandering stream in north central North Dakota, Saskatchewan, and Manitoba. Its basin is, in many ways, a perfect subset or miniature of the overall region, right down to its drainage, glaciation, and flatness (see figure 6.1). The river flows through a basin that is the bed of the extinct glacial Lake Souris. On the Canadian side of the border, the Souris begins between Reyburn and Regina in Saskatchewan. Its middle portion dips into the United States for a significant distance, and then curves back north into Canada and enters the Assiniboine River just southeast of Brandon, Manitoba. The wetlands that line the river's valley and basin naturally slow down and filter the water coming through the system. The resultant slow flow is environmentally essential for water quality and the health of the river system, but it can hinder the quick passage of excessive waters during flood events. The river's path is mostly the result of regional run-off and snowmelt, and so is characterized by heavy spring flood peaks and often very low summer flows.

In the 1950s and 1960s, significant flooding along the Souris in both Canada and the United States culminated in some disastrous episodes. Minot, North Dakota, Estevan and Reyburn, Saskatchewan, and other towns along the Souris received heavy damage in 1969 (United States 1971). As a result, officials in the United States and Canada decided to begin to discuss a means of controlling the flood peaks of the Souris River on both sides of the border. Though the Souris River Basin flood control project had two stated goals, flood control and water supply assurance, flood control alone became the priority. Other results, such as water supply assurances, wildlife refuges, and recreation, were to become secondary, though they were

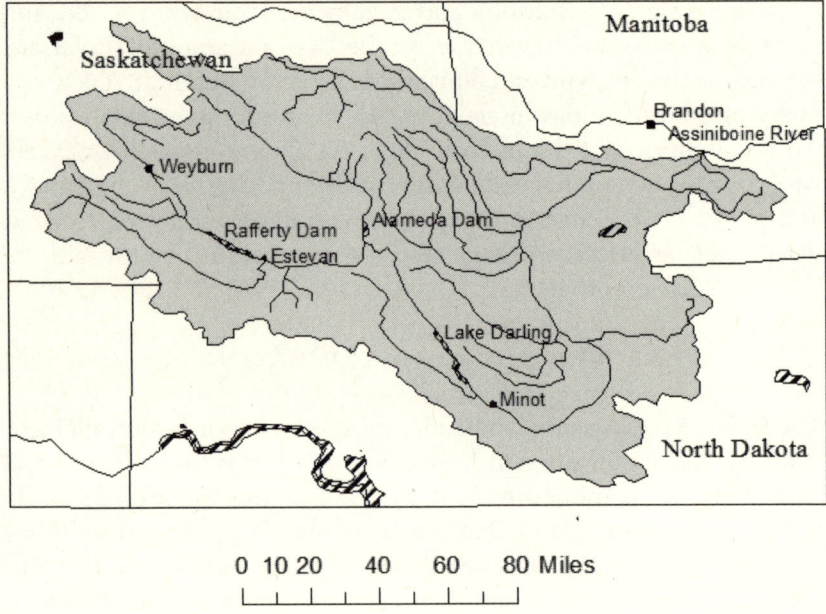

Figure 6.1 The Souris River Basin (Paul Sando – MSUM, Data ISRB)

continually discussed. The issue of water supply assurance has returned to primary importance in the last decade with the rise of the Northwest Area Water Supply Project (NAWS).

Three dams on the Souris (one on a major tributary) are the main portions of the project. In Saskatchewan, the Rafferty and Alameda dams impound and control the release of the spring run-off. In north central North Dakota, the Lake Darling dam not only provides flood control for communities in the United States, but also controls the river downstream in Manitoba. The three dams operate in concert to provide a significant amount of impound space for water and controlled release. The International Souris River Board, which meets annually, manages the cooperative operation and decides on action plans for the basin.

The result is a project that works to this day and does exactly what it was intended to do. Moreover, officials on both sides of the border hailed the project as a model of Canada-US and province-state cooperation and good relations. The floods of the extremely wet winter and spring in 2011 tested the limits of the system's design, and proved that mother nature can overwhelm even the best system.

However, the system proved its worth as recently as the 2009 and 2010 flood seasons, which were disastrous for other parts of the region but were controlled in the Souris Basin.

Into this success comes a separate project, the NAWS pipeline, which has the potential to disrupt the cooperation built over the years by the flood control efforts. The NAWS pipeline is connected to the Missouri River Basin's Pick-Sloan project, a much larger water project in the United States (United States 2009). When the Garrison dam on the Missouri River was completed in 1953 and Lake Sakakawea formed, the act that created the Pick-Sloan project stipulated that North Dakota was entitled to a share of the water impounded as recompense for the lands flooded by the reservoir. NAWS proposed to tap into that water, potentially only using a small amount of what flows through the reservoir in a given season. The water would be piped to a water treatment facility in Minot, North Dakota, to provide a backup source of water for Minot and, more importantly, primary water supplies for the several rural water districts that are a part of NAWS. As a result, the rural districts would not need to construct their own treatment facilities and would have a replacement water supply for their poor groundwater reserves. In theory, the proposal would not only provide an adequate water supply, but also save a considerable amount of money.

The conflict arises between US and Canadian officials and public over questions of environmental policy, in that the water crosses not only a basin divide but also a continental divide as it moves from the Missouri basin to the Souris basin. The Canadian provincial and federal governments immediately voiced concern over cross-basin contamination by biota, should anything happen to the large two-metre diameter pipe that brings water from Lake Sakakawea to Minot. Downstream interests were also concerned about the project's withdrawal of water from the Missouri system. In the last decade, the conflict has grown acrimonious and is perceived as a threat to the highly touted cooperation the International Souris Basin Commission created. NAWS supporters see delays and higher costs due to the legal and political manoeuvring. One proposal to address Canadian concerns is to treat the water before or during its progress in the pipe to eliminate the possibility of contamination before it crosses the boundary. This solution would mean greater cost, as a facility would have to be built to do this and would essentially duplicate the function of the Minot treatment facility. At this point, the NAWS pipe is in place and the initial lines to the rural areas are in place and in use (Minot provides the water out of its supply). No water has yet flowed cross-basin from the Missouri.

THE BE-DEVILLING LAKE

The Devils Lake basin is a geographic and glacial fluke, a basin that has been separated from the Red River and Hudson drainage. As the last glaciers retreated, a large block or several blocks of ice broke away and became buried in the sediments deposited at the foot of the glacier (see figure 6.3) (United States Geological Survey 2009). As this ice melted, it created the deeper portions of the basin. Its northern areas are marshes, wetlands, and streams that drain south into the lake. The moraine deposits that make up the southern border of the Devils Lake basin trap the water in the lake, so that the only natural outlet is groundwater seepage, though there is also a small man-made outlet. In recent years, the lake has grown to immense proportions and has flooded valuable lands and threatened the communities on its shores. Recently, the lake has begun to threaten the region in two directions. If it spills to the north, there is the possibility that excess water will drain cross-basin and into other smaller rivers. If it creates its own natural outlet and spills, uncontrolled, into the Sheyenne River, it will potentially threaten everything and everyone downstream. The Devils Lake basin is indirectly part of the larger Red River basin, and so the problems of run-off from the lake also affect Canadian interests (see figure 6.2).

Past projects to drain the marshes and wetlands to the north for farmland have exacerbated the run-off into the basin. This additional run-off has pushed the lake to near-critical levels. The obvious solution is to build a means to control its level, and a first outlet, completed in 2007 and 2008, pumps water out the southwest corner of the lake into the Sheyenne River. Unfortunately, the wrangling over where the water would go and what pollutants would have to be removed from it resulted in the construction of a very small outlet that has little hope of keeping up with the inflow of run-off. Proposals to increase the size of the existing outlet or build larger ones have resulted in conflicts between American and Canadian interests. The argument spilled over into a seemingly unrelated issue, as North Dakota and Manitoba officials began to accuse each other of blocking efforts to

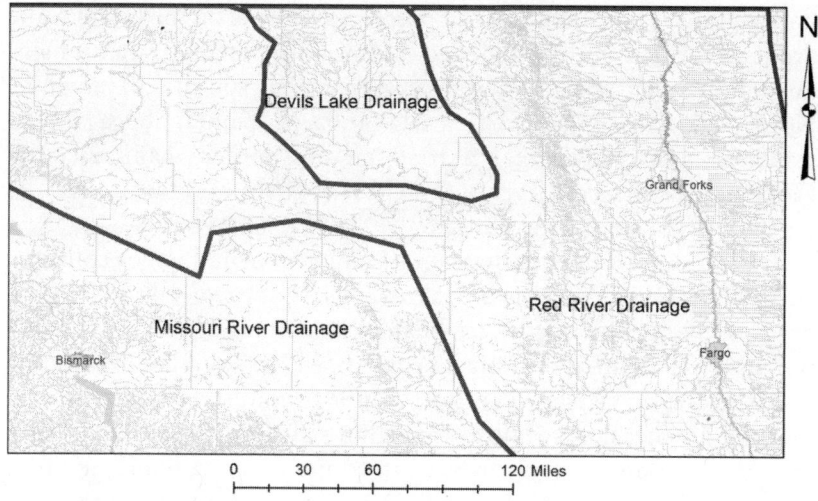

Figure 6.2 Drainage context of Devils Lake (Paul Sando – MSUM, Data ESRI)

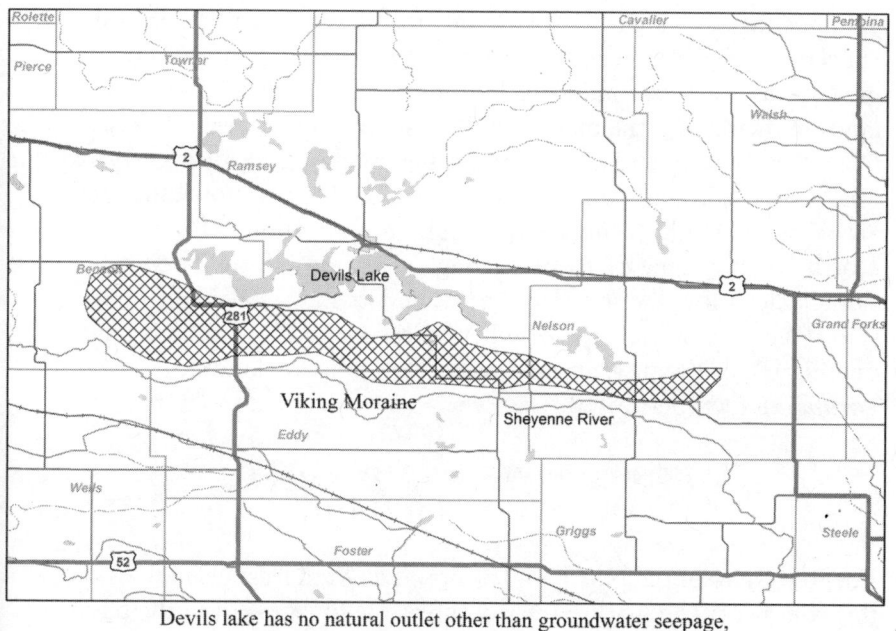

Devils lake has no natural outlet other than groundwater seepage, any southward flow is blocked by the Viking Moraine.

6.3 The glacial geology of the Devils Lake basin (United States Geological Survey basemap and data, State basemap 1963, revised by the author)

find solutions. The Canadians opposed movements on the Devils Lake drainage for reasons that included dealing with potential environmental pollution by non-point source agricultural run-off. Meanwhile, North Dakota officials accused Manitoba of maintaining a double standard. They allege that a road on the border, maintained by Canada, was in fact a levee that held Red River floodwaters on the North Dakota side of the border.

At the moment, negotiations on the environmental issues continue with some progress. North Dakota has proposed a larger outlet drain on either the northwest or northeast sides of the lake where pollution issues are not as big a problem, which would address some of Manitoba's concerns. The State Water Commission has proposed that the northeast drain flow through a paleodrainage channel and into Black Slough to the northeast in conjunction with a large funding request from the state, and has argued this could be completed relatively quickly (Sando 2010a; *Fargo Forum*, 28 October 2010). Time is now the problem. The lake is within just six feet of spilling uncontrolled down the Tolna Coulee to the southeast and into the Sheyenne River. As recent years have been wetter than normal and predictions for wet conditions continue, all remaining prairie wetlands in the region (the buffer for Devils Lake) are either full or near to it (Sando 2010a). North Dakota has made provisions to try to "harden" the Tolna Coulee to accept a possible overflow, but such an overflow would be uncontrolled and a poor option. The state of North Dakota appears to be driving the action led by its congressional delegation, the state legislature, and its state engineer, who has said, "We cannot wait to see where the Federal Government lands on this issue, because at this point they are not landing" (*Fargo Forum*, 28 October 2010).

THE RED RIVER OF THE NORTH AND FLOOD CONTROL

The Red River of the North is also a product of the last glaciations on this continent. The mature river occupies the lowest spot in the basin – once the lake bed of the glacial lake Agassiz – and is the primary stream that flows north into Lake Winnipeg, the Nelson River, and eventually Hudson Bay. In the past one hundred years, the Red River Valley has been the scene of some truly enormous flooding, which culminated in two record-breaking floods in 2009 and 2010. In 2009, the Red experienced a flood the likes of which could only be compared to one that devastated the region in 1895. The record 2009 crest of 40.8 feet – a full 22.8 feet above an 18-foot flood stage at Fargo, North Dakota, and Moorhead, Minnesota – was barely contained by the efforts of those communities, and the downstream effects were just as serious. The 2010 crest reached 37.5 feet, and left a relatively comfortable margin of security as the levees were built to 44 feet. This prompted much discussion of how to control the flooding and reduce the impact on the area. The potential loss if the flood control efforts were to fail, especially in the Fargo-Moorhead area, would be huge economically. Eyes quickly turned north for examples of possible solutions. In Grand Forks, North Dakota, notorious for the disastrous flood of 1997, there are the examples of floodwalls and new green space for water impoundment. In Manitoba, the Winnipeg floodway directs the Red around the city in times of high flow. In the initial rush of discussion, the talks centred on a diversion of floodwaters such as those devised in the Winnipeg floodway. The initial US Army Corps of Engineers studies boiled down to four possible diversions that ranged in size from 20,000 cubic feet per second (cfs) to 35,000 cfs. Two of the proposed diversions were on the Minnesota side of the river, and two were on the North Dakota side. After much wrangling with land, eminent domain issues, and estimates of the size necessary to control the flood, the preliminary findings pointed to a 35,000 cfs diversion on the North Dakota side of the river (see figure 6.4).

Then the real arguments started. Large projects run by the government under the auspices of the Army Corps of Engineers often have a real political and cost handicap, and are frequently poorly perceived among many special interests in the region. There was also the immediate complaint that a rush to a fix might do more harm than good. The opposing view is that delay may mean disaster for the major regional population, health, and business centre of Fargo-Moorhead.

Downstream communities to the north were the first to weigh in with concerns about the impact that 35,000 cfs of water would have on them. In most estimates, the diversion would add up to ten inches on a flood crest downstream (Gunderson 2010). This would have a potentially enormous impact on many small communities along the river. Farther downstream, officials in Grand Forks and Winnipeg remained initially reticent. This silence was initially assumed to be watching and waiting as those communities already have flood control projects in place.

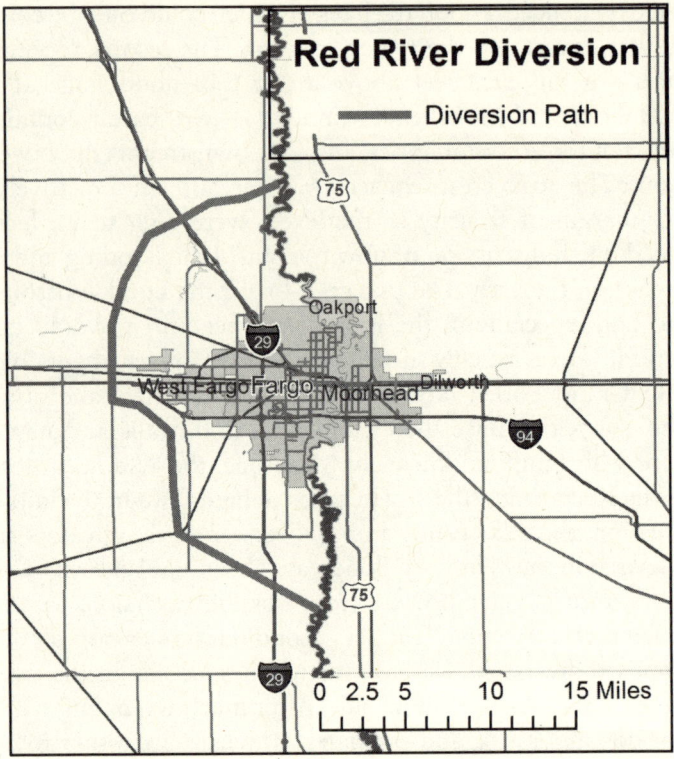

Figure 6.4 Proposed Red River diversion (Paul Sando – MSUM; Data ESRI, US Army)

Further, it was not certain whether or not it would be possible to control the run-off once it began. Many farms in the Red River Valley have added drain tile or other means of speeding drainage to their fields, and many officials and members of the public wondered whether these methods increased the spring flow of the Red. At first glance, the anecdotal answer seemed to be that to speed up runoff would increase the flow into the river and potentially increase ongoing flooding. After reflection it became apparent that the answer was not as simple as it first appeared. One related measure proposed in public forums was to see whether retention methods, such as leaving some water on the fields or wetland redevelopment using dams, would significantly reduce flooding. Several experimental wetland projects have been ongoing in Minnesota and North Dakota for a number of years, and restoration is seen as a potential template for run-off retention and creation of wildlife management areas (Schultz 2001). Another proposal – called the "waffle" proposal because the fields would act as storage units for water like the cells of a waffle store syrup – required paying landowners to retain water on farmland in the spring and release it slowly (Northern Rockies and Prairie Regional Center 2010). Such a proposal would require an expensive program like a conservation reserve project and might disrupt or delay the all-important spring planting season. There is also controversy over whether or not a waffle proposal would actually reduce flooding. Finally, there are proposals by the state and corps of engineers for more dams on tributaries of the Red. One dam, the Maple River dam, is already in place and appears to be functioning exactly as proposed. It has created a temporary flood pool that only fills when there is excess run-off, which it then slowly releases. At all other times the Maple River flows normally through the dam. There are now proposals afoot to place more small dams like these on other tributaries of the Red.

As of this writing, the Red River diversion, additional dams, or other retention methods are being intensely studied and hotly debated in public forums and state government. Many people still consider the planned major diversion around the Fargo-Moorhead metro area to the west to be the ultimate solution, but it is not yet a sure thing. The future is up in the air.

COMMONALITIES OF CONFLICT, AND CONCLUSIONS

Three aspects define each project: complexity, environmental concerns, and geography. All three areas of flooding could see similar results if they have similar configurations based on these three commonalities.

Complexity

When a project has a drive for simplicity inherent in its goals, it often seems that success follows. The Souris River Basin flood control was initially meant to be just that: flood control. The project was conducted in an area that is less complex than others in the region in terms of human occupation and activity. Multiple parties agreed on a course of action and the result simply works. In recent years, the addition of NAWS in the Souris Basin has made the basin projects' intended tasks more complex, with predictably messier results.

The Devils Lake drainage issue is a human-induced one, as it resulted from projects that eliminated wetlands and streamlined drainage into the lake. The lake then over flowed and encroached on occupied lands and agricultural areas. However, draining the lake is problematic since the water would affect areas downstream in the Sheyenne and the Red Rivers.

Had Red River flood control been attempted years ago, such as was done with the Sheyenne River bypass around West Fargo, North Dakota, it may have been easier to complete and more successful. For one thing, more wetlands might remain on both sides of the river to add into a comprehensive flood control program. In general, urban growth in the Red River Valley has only complicated the situation. Recent rules changes in regards to eminent domain have also made it more difficult to secure the necessary land for these types of projects.

Environmental Concerns on the Increase

Environmental concern and accountability needs have increased over time. In a recent conversation, Todd Sando (2010b), the new state engineer for North Dakota, affirmed that project managers now have to take into account many things that they did not less than two decades ago: "If you'd told me when I started with the state that we'd have to be considering environmental issues across an international border for a state project, I wouldn't have believed you."

The Devils Lake drainage project has seen an increase in environmental complexity over time as the lake rises, floods more land, and becomes more polluted. NAWS could potentially open up the Souris Basin to cross-basin contamination and pollution. In the past, such considerations would have been treated as secondary. For example, the creation of large projects like the Pick-Sloan project on the Missouri would have had less potential impact from agricultural run-off. Back in 1944 when the project began, agricultural run-off was less chemically polluted and environmental awareness was not as high as it is today.

More groups with more diverse interests and impacts, tribal groups in particular, have seen the collective power of their voices in modern projects increase compared to those of the past. There has also been a change in the method of planning for the whole region.

In the past, the good of the region as a whole often came at the expense of smaller interests. In particular, the Pick-Sloan project ran roughshod over the interests of the Three Affiliated Tribes and extensively flooded their lands. Recent changes in politics, economics, and awareness would have made the Pick-Sloan project all but impossible today. The public, media, and officials on all sides of the issue, including the North Dakota state engineer, argue this is the result of more environmental consciousness and the stronger assertion of tribal rights (Sando 2010a, 2010b; *Fargo Forum*, 28 October 2010; Conan 2010). Others, such as the United States Geological Survey or joint boards like that of the Souris Basin, point out that changes to the legal systems on both sides of the international border have driven the change of philosophy (Sando 2010a, 2010b; International Joint Commission 2002; International Souris River Board 2007). Avoiding legal fights is always cost-effective.

Geography: Everything Flows North

One thing can be said of all three projects: everything eventually drains north. That geographic fact gives Canada considerable moral and legal leverage in its interests towards these projects. In addition, all of these projects, at least at this point in time, deal with water in excess. There are different rules for water scarcity (Schindler 2010). The controlled movement of excess water through a region in a way that does not adversely impact the interests of the many groups present is a tall order.

In adhering to the concepts of riparian rights and responsibilities, the effects on others are important. Drainage implies that water will move somewhere, through someone else's property or interests. In the Devils Lake drainage project, for example, Canada is concerned with increased run-off and pollutants downstream, the Spirit Lake tribal organization is concerned with moving water south out of Devils Lake and across its lands, and landowners and communities in the Sheyenne and Red River Valleys are concerned with excess water and possible pollutants crossing their lands. If the Devils Lake water were to drain north, bypassing the Sheyenne River and Upper Red River, there would likely be another set of interests to consider as it enter still more streams on the way to the Lower Red River. Canada, through its international legal power, is the region's ultimate downstream interest.

CONCLUDING THOUGHTS OF CONFLICTING INTERESTS

Controlling and mitigating the dual effects of excess water and the imposition of environmental protectionism at the expense of others are not easy issues to resolve.

Controlling run-off is especially difficult in an area where human engineering has thoroughly altered the natural geography. Both the dry years of the 1930s and the wetter periods of the most recent decade have determined the region's priorities. There has also been a net loss of up to 90 per cent of natural wetlands, which are essential in slowing run-off in the region. Project proposals have more often than not modified their plans in mid-application and there is often little data to assess the efficacy of projects because they seldom resemble their proposals. Once run-off is in river systems, managing the river becomes an issue. Levees, diversions, and dams have all been used on the rivers of the area to greater or lesser success. In effect, the mitigation of excess water is perhaps easier to deal with after it becomes mainly an engineering problem.

The socio-economic and political pressures that lead to environmental protectionism are more difficult to manage. Protectionism in this sense is not entirely about physically protecting an environment, but is rather about maintaining a status quo that benefits one party or another in a given area. For example, downstream interests have been content to let the Devils Lake problem run its course so long as it does not alter their immediate environment. If flood stage waters build in the American part of the river, it serves Canadian interests insofar as it delays release into the Canadian part of the system. As the lake nears a crisis that could affect downstream interests, international legal action seems imminent. Unfortunately, the time has come for all parties to realize that they are part of the same multifaceted problem. In a response to criticism that the Red River diversion plan did not take this into account and failed to properly assess downstream effects, Craig Evans of the St Paul office of the US Army Corps of Engineers insisted that his organization had not yet completed its assessment: "We haven't determined where it stops. We haven't found the endpoint of the damages" (Greenwire 2010).

Cooperation between all parties, like that achieved with the Souris River Basin projects, is now a necessity rather than a goal. Such cooperation requires vision and hard work, which are not in short supply, and time, which may very well be. In the case of at least two of the three projects, the results remain to be seen. On 18 January 2011, the National Weather Service released the hydrologic forecast for the Devils Lake basin and the Red River Valley of the North, which predicted Devils Lake would see an additional two feet of water on top of its already-record level. There was a 25 per cent chance that the Red River would meet or exceed its record crest of 2009. In the summer of 2011, the Souris River experienced its largest floods since 1969 – the result not of a spring flood the system was designed to handle, but of extreme precipitation. The winter of 2011 and 2012 was an open winter in the Red River Basin, and flooding was not an issue that spring. By late 2012, the lake had stabilized at just below record level as a result of a dry summer and the opening of a second drain into the Sheyenne River via the Tolna Coulee. However, as the general climate of the region seems to be shifting to a generally wet cycle, time in which to address water issues is indeed in short supply.

REFERENCES

Associated Press. 2010. "Fargo's Long-Term Flood Solution: 10 Years, $1.3B." 22 March.
Conan, Neil. 2010. "A Discussion of Red River Flooding." *All Things Considered*, National Public Radio, 17 March.
Fargo Forum. "$4.2 Million for Devils Lake Flood Relief." 2010. *Fargo Forum*, 28 October.
Greenwire. 2010. "Water: Flood Fears Downstream Hinder Plans to Divert Red River of the North." 27 August.
Gunderson, D. 2010. "Opposition to Red River Diversion Project Grows." Minnesota Public Radio, 9 August.
International Joint Commission. 2002. *Cross Border Cooperation: The Souris River Basin and Flood Control*. Distributed by Prairie Public Broadcasting.
International Souris River Board. 2007. *Forty-Ninth Annual Report to the International Joint Commission Covering Calendar Year 2007*. Regina and Bismarck: The Board.
Northern Rockies and Prairie Regional Center. 2010. "Red River Factsheet, March." Missoula, Montana: The Center.

Sando, Todd. 2010a. "Devils Lake Flooding Disaster: A Stump Lake Perspective." Testimony Presented to the [North Dakota] Senate Budget Committee, 27 August.

– 2010b. Interview with the author. August.

Schindler, D.W. 2010. "Potential Problems with Cross-Border Water Issues." *Canada and the New American Empire.* 22 September. http://www.globalcentres.org/can-us/water_schindler%26hurley.pdf.

Schultz, Steven D., and Jay A. Leitch. 2001. "The Feasibility of Wetland Restoration to Reduce Flooding in the Red River Valley: A Case Study of the Maple River Watershed in North Dakota." *Agribusiness & Applied Economics Report No. 432a.* October.

Simonovic, Slobodan, and Kristine Juliano. 1999. *The Impact of Wetlands on Flood Control in the Red River Valley of Manitoba.* A Final Report to International Joint Commission, September.

Stender, Carol. 2010. "Red River Flood Control Proposal Sparks Intense Interest among Landowners." 14 October. http://www.agrinews.com.

United States. 1971. Army Corps of Engineers. *The Mouse that Roared: The Souris River Flood of 1969.* Distributed by the US Army Corps of Engineers.

– 2009. Department of the Interior. *Record of Decision for the Northwest Area Water Supply Project and Final Environmental Impact Statement on Water Treatment.* 15 January.

United States and Canada. 1989. *Boundary Waters: Agreement between the Government of Canada and the Government of the United States of America for Water Supply and Flood Control in the Souris River Basin (with Annexes and Canada/Saskatchewan Agreement).* Treaty Series 1989 no. 36.

United States Geological Survey. 2009. North Dakota Water Science Center. "Geology of the Devils Lake Basin." December. http://nd.water.usgs.gov/devilslake/science/geology.html.

THE NATIVE BORDER

7

Border Studies and Indigenous Peoples: Reconsidering Our Approach

ZALFA FEGHALI

In political geography, [a border is] an imaginary line between two nations, separating the imaginary rights of one from the imaginary rights of the other.

Ambrose Bierce, 1985, 18

[Some] boundaries are mere lines drawn upon the water, often disrupted or even erased altogether by the lived experiences of First Nations people.

Karl S. Hele, 2008, xi

Ambrose Bierce's above definition of a boundary, taken from his satirical *The Devil's Dictionary*, is an appropriate entry point for my discussion. Over the course of this essay I will identify some drawbacks of ignoring the relationships between border studies, hemispheric studies, and nation, particularly in reference to First Peoples in Canada and the United States. I will make a case for a broader, more comprehensive understanding of and approach to border – and by extension hemispheric – studies in North America. Most conceptions of such studies tend to focus primarily on the relationships between the United States and Latin American states, especially Mexico (though Mexico is a North American state). This is even more the case in studies of the United States' land borders; only recently has the focus shifted from what Gloria Anzaldúa (1999, 25) famously describes as the "open wound" that is the Mexico-US border to the 3,400 kilometre "longest unsecured border" between the United States and Canada. As such, in this essay I work from the need to recognize what Albert Braz (2005, 80) describes as "the urgency of placing inter-American studies in a truly continental context; that is, of bringing the north of America into America."

In my argument for more representative approaches to border and hemispheric studies, I will pay special attention to the First Peoples for whom the international political boundaries between Mexico, the United States, and Canada signify a different challenge, not least because for them, perspectives on North American borders are far more complex. This difference in perception is partially rooted in legal precedence: the Jay Treaty of 1794, for example, included the crucial implications that these groups were not required to adhere to United States and (then Great Britain) Canadian border control and customs regulation. The Jay Treaty also confirmed indigenous rights to transport goods, such as tobacco, alcohol, or other American products, into Canada. Of course, given recent changes in security protocol since 11 September 2001, both the United States and Canada are attempting to circumvent this integral provision in the name of homeland security. Certainly more work needs to be done on how these issues affect legal agreements with First Peoples.

With the exception of a few studies,[1] most scholarship has examined the Canada-US border using a comparative approach that re-inscribes the long-established habit of considering the Mexico-US border (and scholarship on it) as *the* representative border from which border studies as a whole has risen. This is not an altogether invalid approach; there are strong arguments to be made about the continued significance of the Mexico-US boundary in border studies discourse. That said, there are equally compelling arguments (some of which I allude to here) in favour of reconsidering the ways in which we study borders in North America. Of crucial significance in this essay will be questions of empire and nationalism, which many scholars use to characterize the nature of the United States' – and more recently, Canada's – relationship with and perception of borders. Indeed, thinking about empire and nationalism in North America does not require a drastic leap of the imagination to come to the issue of land rights and the more recently pressing issue of border security.

Edward Said's remarks about the nature of travelling theory connect particularly well to the issues of borders I raise in this essay.

According to Said, the theories that emerge from human experiences are solidly rooted in historical moments. As such, these theories are intense and powerful representations of very specific circumstances. They become relatively tame once time passes and situations change. Yet, according to Said (2001, 451), even a tamed theory can regain its influence, since the act and process of "figuring out where the theory went" and remembering what it is capable of allow "its fiery core [to be] reignited," which revitalizes it and makes it relevant once again. I use Said's theory as a primary rationale to begin analysis at the Mexico-US border, before I discuss Gloria Anzaldúa's work and assess whether her theories travel well and can be transposed to Canada-US border analyses.

To frame this discussion, I turn to Pablo Vila, an anthropologist of the Mexico-US border. He writes: "[Some] authors tend to homogenize the border, as if there were only one border identity, border culture, or process of hybridization. I think, instead, that the reality of the border ... goes well beyond that consecrated figure of border studies, the border crosser" (2003, 321). While Vila is talking about the Mexico-US border, he could very well be referring to the Canada-US border or an "invisible" border such as that of class. Yet here it is necessary to support the "necessity of generating discursive modes originating primarily from the native or indigenous cultural context, as it informs Native American literary texts, and suggests ways in which such discursive strategies can be articulated" (Pulitano 2003, 60). Indeed, as Jesse Peters (2005, 337) notes, "It is important to recognize the fact that both Euroamerican and Native ways of knowing are necessarily bound together." In fact, I would argue, as Pulitano (2003, 17) does, that it is absolutely crucial to weave "Native American forms of discourse into the rhetorical patterns of the classical tradition. Native American theorists and writers invite all of us to open ourselves to new ways of being in the world, ways that are very different from the models that we have been given by the Western hermeneutical tradition." Indeed, "any comparative approach to border tribes along US boundaries ... needs to take into account each tribal nation's differential historical relationship to the nation state in which it now resides, relationships that are often codified in treaties" (Sadowski-Smith 2008, 95).

As I note above, it is now common knowledge that Canada-US border is often sidelined in favour of its counterpart to the south. While Anzaldúa (1999, 25) characterizes the Mexico-US border as "*una herida abierta* [an open wound]," a site at which "the Third world grates against the first and bleeds, [creating] a third culture – a border culture," the Canada-US border has only recently been popularly perceived as a comparable case. Critical materials that address the Mexico-US border could benefit a reading of the Canada-US border, particularly in the context of First Peoples. Indeed, there are potentially useful and transferable concepts that can help build a more comprehensive border studies in the Americas. That said, it is wise to remain cognizant of the complexities and potential dangers of this enterprise, for as Konrad and Nicol (2008, 56) warn, "Those who study borders, and attempt to theorize borders, must ever be vigilant to the need to consider borders and borderlands human and social constructs in transition, and to avoid containment in thinking in the silos of culture, nation-state and identity."

Summer 2010 witnessed what many consider to be the most recently popularized event in First Peoples' struggle for recognition (both in the United States and internationally) as nations. In July 2010, the United Kingdom declined to issue visas to the Iroquois Nation's lacrosse team, which had been due to compete in an international competition, since Iroquois passports are not currently recognized as state-issued travel documents. The United Kingdom clarified that it would waive the visa requirement and accept the Iroquois document if it was "accompanied by a United States passport" (BBC, 15 July 2010). This was despite Secretary of State Hillary Clinton's "one-time" letter of assurance to the UK government that team members would be allowed back into the United States on their tribal passports. Clinton's unintentionally ironic guarantee that the lacrosse team would be allowed back into the United States sparked a controversy that eventually led the US State Department to confirm in a statement that it had no treaty obligation with the Iroquois nation to recognize its passports, despite the Iroquois "status" as a nation. Unsettling as it remains, this incident reflects the larger issues connected with the recognition of native/First Peoples as nations in more than name only, and brings the question of citizenship into play. Indeed, this is not the only example of "struggles for sovereignty [which] have found expression in maintain[ing] Iroquois border crossing rights, which include the right to work, hunt, fish, and trap on either side; to carry sacred objects and personal goods without paying duties" (Sakowski-Smith 2008, 88). Sadowski-Smith correctly identifies another layer of complexity here, as "tribes like the

Tohono O'odham [whose lands straddle the Mexico-US border], for example, have demanded that US citizenship be granted to enrolled tribal members living in Mexico," while the Iroquois "insist on recognizing that their membership in an independent nation is uniquely defined by international treaties between the United States, Britain, and Canada" (95).

For the purpose of this essay, I borrow T.H. Marshall's (1964, 91) definition of citizenship, which he sees as "a direct sense of community membership based on loyalty to a civilization which is a common possession." Here, Marshall presents a causal relationship between membership in a community, civilization, and citizenship. This concept of citizenship is also contingent on a certain understanding of the modern nation state, which as Maria-Barbara Watson-Franke (2009, 104) puts it, "encompasses the notions of the people as a nation, the sovereignty of these people as a nation, and the state as the sum of its individualised citizens." In these views, the citizen is a political entity and the smallest unit that makes up the state. As such, the citizen does not only enjoy the rights that come with membership in the political community, but also assumes the duties. In this framework, citizenship defines identity (Joseph 1999, 163). However, one relationship that Marshall's definition does not address is that of territory and citizenship. How does this conception play out in the context of First Peoples' lived realities? According to article 73 of the Iroquois Confederation's Constitution, for example, "The soil of the earth from one end of the land to the other is the property of the people who inhabit it. By birthright the Ongwehonweh (Original beings) are the owners of the soil which they own and occupy and none other may hold it. The same law has been held from the oldest times" (National Public Telecomputing Network and the Constitution Society n.d.). In this view, for the Iroquois it is not state-issued citizenship that has the greatest hold, but the attachment and experience of living on and using land and territory. Indeed, the Iroquois Confederacy, which is comprised of "Mohawk, Oneida, Onondaga, Cayuga, Seneca, and Tuscarora Nations, connects communities with similar languages and relationships to a specific place – the land and region encompassing Lake Ontario and Lake Erie" (Sadowski-Smith 2008, 88).

The Mohawk Nation of Akwesasne is one of two tribal territories, along with the Míkmaq of Maine, that physically straddle the Canada-US border. The international border as it stands today was established in part as a result of treaties between the United States and the United Kingdom, specifically, the Jay Treaty (1794), the Treaty of Ghent (1814), and the Treaty of Washington (1836). While these agreements were not signed with First Peoples groups, as I have shown above, the Jay Treaty in particular demonstrates "First Nations were an essential element in diplomatic relations between these two governments. This Treaty between two European Nations acknowledged that protection of First Nations' rights [was] an important part of the non-Indian reality" (Aboriginal Rights and Research Office 1999).

To the south, four Native American nations have tribal lands that straddle the Mexico-US border: the Tohono O'odham, the Yaqui, the Cocopah, and the Kickapoo. The Tohono O'odham nation, with a population of approximately 22,000, is recognized by the US federal government as a First Nations group. Its reservation lands comprise nearly 3 million acres in southern Arizona and its traditional tribal lands extend south into the Sonoran desert in Mexico. The Yaqui nation has reservation lands of about 1,000 acres in New Pascua, Southwest Tucson, and southern Arizona, and obtained US federal recognition in 1978. The Cocopah has reservation lands of 6,000 acres and a population of 4,000, approximately half of which resides in the Colorado River delta region of Mexico, while the US part of the tribe is recognized by the US federal government. The Kickapoo nation comprises about 600 people, with a 125-acre reservation in Maverick County, Texas. According to Amnesty International (1998), though the Treaty of Guadalupe-Hidalgo recognized Native American tribes' rights as sovereign nations to cross the new border "without hindrance, human rights monitors in the Arizona region have documented instances in which Native Americans who wish to cross the border to visit family and attend ceremonies have been harassed and had problems complying with the documentation required by Immigration and Naturalization Services."

To be sure, the attacks of 9/11 changed the reality of and discourse around borders, particularly given figures in the United States and Canada such as Brigitte Gabriel, who urges Americans and Canadians to stay vigilant because "the terrorists are using our borders to infiltrate our country" (quoted in Leclaire 2010). Because of them and because of legislation such as the Secure Fence Act of 2006, which doubled the funding for border patrol agents in the United States, the American and Canadian borders have become key arenas in which the "War on Terror" is played out. The implications of this development on First Peoples' lived experiences, while downplayed or not addressed in border studies scholarship, should not be ignored.

Among the more controversial changes in border security that took place after 9/11 was the Real ID Act of 2005. Under the act, "notwithstanding any other provision of law, the Secretary of Homeland Security shall have the authority to waive all legal requirements such Secretary, in such Secretary's sole discretion, determines necessary to ensure expeditious construction of the barriers and roads."[2] Using this power, Secretary of Homeland Security Michael Chertoff decided to "waive in their entirety" the Endangered Species Act, the Migratory Bird Treaty Act, the National Environmental Policy Act, the Coastal Zone Management Act, the Clean Water Act, the Clean Air Act, and the National Historic Preservation Act in order to extend triple fencing through the Tijuana River National Estuarine Research Reserve near San Diego. The Real ID Act further stipulates that his decisions are not subject to judicial review, and in December 2005 a federal judge dismissed legal challenges to his decisions by the Sierra Club, the Audubon Society, and others.[3] This triple fencing also affects those First Peoples whose tribal lands legally extend across the Mexico-US border.

Commentators have criticized the United States for failing to address the United States Border Patrol's widespread abuses of indigenous peoples travelling in their own territories, or the violations of Native American Grave Protection and Repatriation Act (NAGPRA) and other federal laws during the construction of the Mexico-US border wall. It would seem that the perceived threat of an open border to the south, which necessitated the construction of the border fence in the name of homeland security in the first place, renders agreements made with cross-border First Peoples communities secondary at best. Again, questions of citizenship come into play.

These questions of citizenship are explicitly tackled by Gloria Anzaldua, who gained a cult following after the publication of her 1987 book *Borderlands/La Frontera: The New Mestiza*. Her work blurs the boundaries of a range of different theoretical approaches and contexts, including postcolonial, feminist, border, and queer theories, in addition to other disciplines, including sociology and anthropology. The versatility and importance of *Borderlands/La Frontera* means it is invoked in scholarly discussions of Chicana/o literature, feminist theory, and border studies, among other fields. Gita Rajan and Radhika Mohanram (1995, 5), for example, note how "recent critical focus has shifted from exile and diaspora to borders, and the crossing and re-crossing of physical, imaginative, linguistic, and cultural borders." They see Anzaldúa's work as "largely responsible for a new direction in postcolonial studies." Within race studies, Anzaldúa's treatment of *mestizaje* (in a culturally specific sense rather than a broader framework sense) has been used in parallel with other processes of racial mixture, such as *métissage* in the case of Canada or other former French colonies. On a broader level, the *mestiza* as a figure of resistance is also used in conjunction with the *métis(se)* and the *mulatta*. Critics such as Rafael Pérez-Torres, José David Saldívar, and Françoise Lionnet insist on the potential of the *mestiza*, as both a literal figure and as a conceptual tool, to instigate change. I further examine this potential below and assess its significance in North American borders discourse.

Said's understanding of the impact of theories that travel across time, space, and disciplinary boundaries is based on the premise that a theory's vigour can be traced to specific historical circumstances. This view encourages an analysis of the historical circumstances from which Anzaldúa's work arose. Indeed, the context from which Anzaldúa writes is perhaps as important as the content of her writing. *Borderlands/La Frontera* was inspired by the Chicano civil rights movement, which began in the 1960s and lasted well into the 1980s. The term *Chicano* refers to a politically engaged individual, most often of Mexican descent, and is adapted from the terms *Mexicano*, *Meschicano*, and *Xicano*. Chicanos generally identify as *Latino*, which means their ancestors were of Hispanic/Latin American ancestry as opposed to Anglo-American/white European descent. The Chicano civil rights movement aimed to gain equal rights as a minority group in the United States. The Chicano Movement, or *El Movimiento Chicano*, is considered to have officially begun in 1965, when César Chávez helped found the National Farm Workers Association. That the first event of the movement was one of farm workers' unionizing reflects the position of Chicanos at that time: the majority either worked in or came from families that worked in agriculture. As either sharecroppers or migrant workers who moved around the United States, these Chicanos did not have equal civil or worker rights and white land or factory owners often discriminated against and exploited them. With the formation of *La Raza Unida* political party in 1970, what began as a farm workers' initiative developed into a fully fledged political movement for civil, social, and economic rights.

In its original meaning, the Spanish term *mestiza/o* refers to the offspring born of the mixing of racial groups. The term *mestizaje* refers to one's *mestiza/o* character – their *mestiza*-ness. It can also refer to the process of becoming *mestiza/o*. The product of interracial mixing, the *mestiza/o* evokes two inextricable issues: projected purity, as seen when *mestiza/o* individuals are considered to be impure, inferior versions of their constituent racial groups; and colonial conquest, which illuminates a silenced history of violence, oppression, and rape. Despite these issues, Anzaldúa is careful to cast her ancestors' original *mestizaje* in a more positive light and focuses on their enhanced ability to survive: "The *mestizos* who were genetically equipped to survive small pox, measles, and typhus, founded a new hybrid race." These were "people of mixed Indian and Spanish blood, a race that had never existed before" (Anzaldúa 1999, 27). This new hybrid people, living on the borders of different racial groups, were the ancestors of many people who today live on the literal border between the United States and Mexico. Indeed, Anzaldúa wrote *Borderlands/La Frontera* from her own experiences of living in this borderland region.

Anzaldúa's *mestiza* exists in borderlands, and is "neither *hispana india negra española/ni gabacha*";[4] rather, she is "*mestiza, mulata,* half-breed/caught in the crossfire between camps/while carrying all five races on [her] back/not knowing which side to turn to, run from" (1999, 216). However, according to Anzaldúa and despite the difficulties engendered by her very existence, the *mestiza* is also a figure of enormous potential as her multiplicity allows a new kind of consciousness to emerge. This *mestiza* consciousness moves beyond the binary relationships and dichotomies that characterize traditional modes of thought and seeks to build bridges between all minority communities to achieve social and political change. Anzaldúa locates what she calls the new *mestiza* consciousness at a site that, as Françoise Lionnet (1995, 5) suggests, "is not a territory staked out by exclusionary practices."

For her part, Anzaldúa presents a series of steps the *mestiza* must take as she moves towards her new consciousness and begins her life of action. The first is to "take inventory," to look at who she is and distinguish which of her parts come from which of her disparate ancestors. This step is important as Anzaldúa makes it a point to expose and reiterate her Spanish *and* "Indian" blood. The new *mestiza* must continue by "putting history through a sieve," enacting a reminder that her own *mestizaje* has its history in conquest, rape, violence, and occupation. After scrutinizing history, in this case Western accounts and justifications of Spanish conquest in the Americas, she "communicates the rupture" with oppressive traditions. In so doing, the new *mestiza* also articulates her own feminist agenda and refuses to remain silent in favour of advancing the male-dominated Chicano cause. As she "documents the struggle" by writing it down, she "reinterprets history," reclaims it as her own narrative, and finally uses "new symbols, shapes new myths" that move away from "Greek myths and the Western Cartesian split point of view." In so doing, the new *mestiza* is able to "root" herself "in the mythological soil and soul of this continent" and thus attains a consciousness that does not adhere to traditional "Western" modes of thought (Anzaldúa 1999, 104).

Anzaldúa's work and the Chicano movement have implications for thinking about the Canada-US border. Obviously there is a vast difference between the two borders, not least because that between Mexico and the United States is partially delineated by a physical wall. Using Said's understanding of travelling theory as my cue, in the remainder of this essay I focus on two common tropes and figures in border studies: conceptions of the border-crosser and the border(lands). I show how both are commonly used in traditional border studies work and then highlight how they could benefit from engagement with native critical approaches before I assess their possible impact on approaches to border and hemispheric studies. Here it is important to remember, as Konrad and Nicol (2008, 96) confirm, "One key to understanding [the borderlands] is to acknowledge the deep history of this interaction zone."

In this vein it is important to clarify that this approach is in no way an attempt to universalize the experiences of all First Peoples (to do so would defeat the very purpose of this exercise). Similarly, it is equally important to note that I do not intend to sideline other borderland concerns with the Mexico-US border either. It should not be the case that studies of North American borders sideline any group at all; hence the need for a broader, hemispheric approach.

According to Claudia Sadowski-Smith (2002, 2), "By the turn of the twenty-first century ... the US-Mexico frontier has evolved into one of the most prominent sites for analyses of border transgressions that emphasize contemporary diasporic practices of hybrid place-making and non-absolutist citizenship." On the other hand, according to Konrad and Nicol (2008, 45), "The Canada-US border was considered as an area of interest only when it became clear that free trade initiatives would have an impact on this rarely acknowledged boundary. This interest, and accompanying commitment to research, has now led to a substantial contribution to knowledge and understanding about the United States-Canada border in its own right." It is easy to trace the reasons for the previous imbalance: the Canada-US border had long been the "longest undefended border in the world" until 9/11.[5] However, as Konrad and Nicol (2008, 2) observe, "The time-word rhetoric about the 'longest undefended border in the world' has disappeared both in Canada and the United States. In its place a new, 'post-9/11' border culture has emerged in the Canada-US borderlands." Understanding this post-9/11 border culture is crucial to analyzing the implications of changes in, for example, border security.

This brings this discussion back, briefly, to Anzaldúa: is the Chicana/o experience universal and truly representative of all border/borderlands existences? Arguably, despite embracing her innate hybrid state and the mixture of Hispanic and indigenous heritage that makes up Anzaldúa's *mestizaje*, her work is not used in discussions of indigenous communities. I hope to come closer to finding out how inclusive, transnational, and/or international Anzaldúa's *mestiza* – as a border crossing subject – really is. What I would like to flag here is exactly this idea that Anzaldúa's border is localized – not only in terms of geography, but also in terms of her own experience – as a lesbian Chicana woman writing in the 1980s in the United States. She has had an enormous influence on several fields in American studies and beyond, and her writing enacted a shift in American studies in the 1980s and 1990s that was, arguably, unrivalled until the attacks on the United States on 9/11. That said, and particularly since 2001, the way borders are approached in North America needs to be reconsidered to provide for concerns about security and globalization.

The contemporary figure of the border crosser is thrown into relief within a First Peoples approach. To consider, for example, Anzaldúa's figure of the new *mestiza* – a figure of hope who symbolizes the affirming potential of border interactions and crossings rather than their prohibitive nature and often violent histories, it is clear that the figure of the border-crosser is crucial to any articulation of border studies.

I take Gerald Vizenor's trickster analysis as my cue and instead of attempting to simply recuperate current approaches to North American studies – a discursive move that both justifies and legitimizes the dominance of "Eurocentric" thinking in this case – I want to advocate for what Vizenor calls indigenous theorizing. Such an approach does not hold that current discourse should be dismissed in its entirety; it merely chooses to highlight the preoccupation of current approaches to border studies that sideline the urgent and pressing concerns of First Peoples. Vizenor, who identifies as Anishinaabe, is also famous for formulating what he has called trickster discourse and advocating for the practice of indigenous literary criticism, which he uses in both his academic and fictional work. It is important to note here that for Vizenor (1988, x), "The trickster is comic nature in a language game, not a real person or 'being' in the ontological sense." Helen Lock (in Vizenor 1990, xiv) notes, "Since for Vizenor the trickster is constituted linguistically, trickster discourse is the process whereby language negotiates the boundaries of the crossblood's world, deconstructing the fixed, authoritative beliefs and definitions that Vizenor has called 'terminal creeds.'" As he puts it, to imagine the trickster we "relume human unities; colonial surveillance, monologues, and racial separation are overturned in discourse" (Vizenor 1988, x). If the trickster is able to turn things on their heads, then doing so is the first step in staking a claim in discourse around North American border studies. While Vizenor does not prescribe steps for becoming a trickster as Anzaldúa does for her new *mestiza* – in fact, to do so would be counter to the very idea of the trickster in discourse – the parallels between the two figures are unmistakable, which of course is not to say that the two figures are the same. However, the figure of the trickster is rarely, if ever, used in border studies, though it fits into the border-crosser paradigm at least as well as Anzaldúa's new *mestiza*. If we are to consider these steps in the framework of a First Peoples identity and border studies, it becomes clear that what Anzaldúa prescribes can be accurately described as *survivance*. As Anzaldúa portrays a postcolonial relationship, First Peoples border communities experience this same dynamic.

To define the concept of survivance, Vizenor (1998, 15) states, "Survivance ... is more than survival, more than endurance or mere response; the stories of survivance are an active presence ... The native stories of survivance are successive and natural estates; survivance is an active repudiation of dominance, tragedy, and victimry." Another question to think about when approaching First Peoples border concerns is whether postcolonial theory provides an adequate vocabulary, or whether a different, more nuanced vocabulary would better address these questions. In the pursuit of a different approach to studying North American borders, it is worth asking, then, what commonalities exist between native and First Nations communities living on and around international political boundaries. How does this affect the way we think about cross border First Peoples in Mexico, the United States, and Canada? Is it enough or even appropriate to say these are exceptions to the rules within border studies? I suggest that these exceptions to border concerns should instead be on the forefront of how these rules are made. Not only has border studies sidelined the Canada-US border until recently – it has hitherto completely ignored the "complications" cross-border First Peoples pose.

Karl S. Hele's comment, my second epigraph, represents another way that borders are perceived by First Peoples people, as "mere lines drawn upon the water," which are undermined by the realities of First Nations' "lived experiences." This also ties in well with Anzaldúa's (1999, 25) understanding of borders as artificial, arbitrary impositions on "the skin of the earth." Crucial here is the understanding of the Canada-US border as a *medicine line* or place of sanctuary from the situation of First Peoples groups in the United States or Canada. Indeed, Sadowski-Smith (2008, 86) notes that this "understanding of the border as sanctuary united various indigenous groups on both sides of the border." Significantly, "they have … provided a home to native people whose perspectives on hemispheric borders between the United States, Mexico, and Canada have remained largely unknown in US border studies and are cited only incidentally in American Indian scholarship" (72). This conception of borders has implications for scholarly engagement with hemispheric and border studies.

Both Anzaldúa's work and Hele's comments highlight the radically different perspectives First Peoples have on borders in North America. It is these views that fuel the impetus to re-conceptualize North American border studies and, ultimately, hemispheric studies. This impetus is rooted in a critique of current North American tendencies to undermine the status of First Peoples groups *as nations*. While in this essay I quite arbitrarily chose to focus on the recent treatment of the Iroquois lacrosse team and provided an overview of treaty violations perpetrated by the United States government against cross-border and other native communities, these are not the only examples of such violations. While I discuss "the reality" of First Peoples' lived experiences, I believe it is important to remember that despite the alternative and often subversive perspectives these groups have on international political boundaries, a large part of their lived experience is as minority groups in larger political and state systems that continue to ignore their rights as nations and undermine the sovereignty guaranteed them by treaties. Using First Peoples tropes and criticism in "mainstream" border studies has the potential to do more than simply make border studies more "representative"; indeed, recognizing the effectiveness and validity of First Peoples' perspectives could have significant implications on other areas. This connects back to my earlier discussion of travelling theory (Said) and citizenship (Marshall), and also asks whether the study of the Canada-US border has implications for the study of the Mexico-US border. Even more broadly, the larger issue is whether and how these perspectives offer new ways to re-conceptualize existing models, paving the way for more "democratic" approaches and simultaneously enriching our understanding of First Peoples.

NOTES

1 Claudia Sadowski-Smith has published several studies on this topic, as have Victor Konrad and Heather N. Nicol.
2 The Real ID Act was passed as a rider on an appropriations bill that funded the wars in Iraq and Afghanistan. The act has seen resistance from environmental groups in addition to civil liberties advocates. For more, see United States (2005).
3 Secretary Chertoff exercised his waiver authority on 1 April 2008. In June 2008, the US Supreme Court declined to hear the appeal of a lower court ruling that upheld the waiver authority in a case filed by the Sierra Club. In September 2008, a federal district court judge in El Paso dismissed a similar lawsuit brought by El Paso County, Texas. See, for example, Johnson-Castro (2008).
4 "[Neither] Hispanic, Indian, Black, Spanish/nor white." My translation.
5 Developments since 9/11 have shifted the description of the border from "longest undefended" to "longest unsecured."

REFERENCES

Aboriginal Rights and Research Office, Mohawk Council of Akwesasne. 1999. "Aboriginal Border Crossing Rights and the Jay Treaty of 1794." http://www.akwesasne.ca/jaytreaty.html.

Amnesty International. 1998. "United States of America: Human Rights Concerns in Border Region with Mexico (including errata)." 18 May. http://www.amnesty.org/en/library/info/AMR51/003/1998.

Alurista. 1981. "Cultural Nationalism and Xicano Literature during the Decade of 1965–1975." *MELUS* 8 (2): 22–34.

Anzaldúa, Gloria. 1999. *Borderlands/La Frontera: The New Mestiza*. Second edition. San Francisco: Aunt Lute Books.

BBC. 2010. "UK Refuses to Grant Visas to Iroquois Lacrosse Team." BBC Online, 15 July. http://www.bbc.co.uk/news/world-us+canada-10634044.

Bierce, Ambrose. 1985. *The Devil's Dictionary*. London: Penguin.

Braz, Albert. 2005. "North of America: Racial Hybridity and Canada's (Non)Place in Inter-American Discourse." *Comparative American Studies* 3 (1): 78–88.

Bost, Suzanne. 2003. *Mulattas and Mestizas: Representing Mixed Identities in the Americas, 1850–2000*. Athens and London: University of Georgia Press.

Fisher Fishkin, Shelley. 2005. "Crossroads of Cultures: The Transnational Turn in American Studies." *American Quarterly* 57 (1): 17–57.

Hele, Karl S., ed. 2008. *Lines Drawn upon the Water: First Nations and the Great Lakes Borders and Borderlands*. Waterloo: Wilfrid Laurier University Press.

Johnson-Castro, Jay J. 2008. "Border Wall Battle: Bad News vs. Good News. Judge Dismisses Lawsuit Against Secretary Michael Chertoff, Filed by the US-Mexico Border Plaintiffs, Ruling in Favor of the Federal Government." 12 September. http://www.narconews.com/Issue54/article3190.html.

Joseph, Suad. 1999. "Women between Nation and the State in Lebanon." In *Between Woman and Nation: Nationalism, Transnational Feminism, and the State*, edited by Cora Kaplan, Norma Alarcon, and Minoo Moallem, 162–81. Durham: Duke University Press.

Konrad, Victor, and Heather N. Nicol. 2008. *Beyond Walls: Re-inventing the Canada-United States Borderlands*. Aldershot and Burlington: Ashgate.

Leclaire, Jennifer. 2010. "Because (Islam) They Hate." *The Voice*. http://www.thevoicemagazine.com/culture/society/brigitte-gabriel-because-they-hate.html.

Lionnet, Francoise. 1995. *Postcolonial Representations: Women, Literature, Identity*. Ithaca and London: Cornell University Press.

Marshall, T.H. 1964. *Class, Citizenship, and Social Development*. Chicago: University of Chicago Press.

National Public Telecomputing Network and the Constitution Society. N.d. *Constitution of the Iroquois Nations: The Great Binding Law, Gayanashagowa*. http://www.constitution.org/cons/iroquois.htm.

Peters, Jesse. 2005. Review of *Toward a Native American Critical Theory*. *The American Indian Quarterly* 29 (1–2): 337–9.

Pulitano, Elvira. 2003. *Toward a Native American Critical Theory*. Lincoln: University of Nebraska Press.

Rajan, Gita, and Rahdhika Mohanram, eds. 1995. "Introduction: Locating Postcoloniality." In *Postcolonial Discourse and Changing Cultural Contexts: Theory and Criticism*, 1–16. Westport and London: Greenwood Press.

Sadowski-Smith, Claudia. 2002. *Globalization on the Line*. New York: Palgrave.

– 2008. *Border Fictions*. New World Studies, edited by A. James Arnold. Charlottesville and London: University of Virginia Press.

Said, Edward. 2001. "Traveling Theory Reconsidered." In *Reflections on Exile and Other Literary and Cultural Essays*, 436–52. London: Granta Books.

Saldívar, José David. 1997. *Border Matters: Remapping American Cultural Studies*. Berkeley: University of California Press.

United States. 2005. Congress. An Act Making Emergency Supplemental Appropriations for Defense, the Global War on Terror, and Tsunami Relief, for the Fiscal Year Ending September 30, 2005, and for Other Purposes. http://www.gpo.gov/fdsys/pkg/PLAW-109publ13/content-detail.html.

Vila, Pablo, ed. 2003. *Ethnography at the Border*. Cultural Studies of the Americas 13. Minneapolis and London: University of Minnesota Press.

Vizenor, Gerald. 1988. *The Trickster of Liberty: Native Heirs to a Wild Baronage (Emergent Literatures)*. Minneapolis: University of Minnesota Press.

– 1990. *Bearheart: The Heirship Chronicles*. Minneapolis: University of Minnesota Press.

– 1998. *Fugitive Poses: Native American Indian Scenes of Absence and Presence*. Lincoln: University of Nebraska Press.

Watson-Franke, Maria-Barbara. 2009. "To Teach 'The Correct Procedure for Love': Matrilineal Cultures and the Nation State." In *The Political Interests of Gender Revisited: Redoing Theory and Research with a Feminist Face*, edited by Anna G. Jónasdóttir and Kathleen B. Jones, 104–21. Manchester: Manchester University Press.

8

Navigating the "Erotic Conversion": Transgression and Sovereignty in Native Literatures of the Northern Plains

JOSHUA D. MINER

Like the Blackfoot characters in Thomas King's *Truth and Bright Water* – like the river, too, that forms a natural boundary between its American and Canadian communities – indigenous border stories tend to ramble. They may head off in difficult directions; and may alternately crash like cataracts or flow like streams, but they orient nonetheless if their process is respected. When the young narrator, Tecumseh,[1] interrupts his aunt to declare that he would "go east," even though he has not heard the first part of her story and does not know its setting (King 1999, 53), he betrays a poor understanding of the way storytelling works for his people. As he comes of age, he will learn how to tell and how to listen. He will learn that well-told stories profoundly ground native cultural identities and lead them in their course.

Tecumseh will also learn that stories and their tellers bear the unique responsibility of enabling *and* raising barriers to mutual understanding between neighbours, for they sustain diverse cultural maps that may resist translation. This paradox is more pronounced for Plains communities near the Canada-US border, where narrative barriers on multiple axes – national, ethnic, ecoregional, local – prevent productive cooperation between indigenous and non-indigenous peoples (McCormack 2005, 110). Here, social and political change is slow. Geographers have begun to understand borders as "one part of the *discursive landscape* of social power, control and governance" (Newman and Paasi 1998, 196); border*lands* are thus complex topographies of discourse construed spatially. Canada-US border spaces are reified by a kaleidoscope of indigenous and nonindigenous cultural practices, complicated by competing constructs like *state/province/territory, reservation/reserve, Indian, Métis,* and *frontier*, which continue to frame the lived experiences of their residents. So, though King (2002, 102) remarks that the political border dividing the native peoples of Canada and the United States is, "after all, a figment of someone else's imagination," it certainly bears real-world sociopolitical, cultural, and psychic consequences.

By referring to the forty-ninth parallel as a figment, King recalls a long history of indigenous understandings of cadastral lines as imaginary and as uniquely powerful for the way Eurowesterners hinge their sociopolitical and cultural identities on them. Centuries before the northern Plains became a thoroughfare for transcontinental railways, native peoples called the Canada-US border the "medicine line" on account of what it could provide them in resistance to these nation-states (LaDow 2001, 41). Many still use the line to their advantage – for example, in the legal limbo of cross-border reservation spaces. Any permanent solutions to conflicts in the borderlands, however, will require that their discursive topographies first be surveyed and then navigated. The situation is distinctly troubling for indigenous people, since story(telling) serves as a locus of discourse by which they have long positioned themselves against colonial empires and neocolonial institutions. Native writers show the possibility to effect change through this narrative dialogic. Here, they may destabilize the national narratives that have subordinated and marginalized them for so long.

As a narrative construct, the *borderland* far exceeds the *border*. In the North American context, it is related to the *frontier*, which connotes a dangerous presence beyond the contested but safe bounds of *territory*, a kind of middle ground between civilization and wilderness. Choctaw-Cherokee scholar Louis Owens (1998, 26) argues that this conception stems from the "ultimate logic of territory," a logic of unremitting "appropriation and occupation" and exclusion, where colonial power draws ever-advancing lines against the unknown and uncontrollable. The Eurowestern eye has always envisioned the American frontier as savage, carnal, and anachronistic. It is a place where the wild "Indian" resisted progress but was either tamed or removed by modernity – in this way, it serves as both adversary and nutriment for territory. Yet borderlands are more interstitial than frontiers, because as narrative spaces they are territorial (in this case, national) *and* on the frontier; they are at once lawed and lawless, civil and barbarian.

Indigenous people embody the borderlands in Canada-US national narratives because they have been discursively constructed as demons secreted beyond nation-states, otherworldly yet dangerously ever-present.[2] This fiction has left little room for any recognition of indigenous political or cultural sovereignty. John Smith and Samuel de Champlain began to articulate the "sauvage" [savage] as early as the sixteenth century and, as expected, their European imaginations and profound misunderstanding of indigenous life helped grow the narrative seeds for colonial fantasies about wild natives. Theirs are the "Indians" that Owens and Anishinaabe author Gerald Vizenor (1998, 32) identify as simulacra born of the colonial imagination, which have no true referent and bear no relation to real indigenous people. Imagined barbarians nevertheless impose themselves on real native experience. Indian savages parade through federal and state legislation, newspapers, journals, sermons, novels, and films, helping to reproduce and validate each other. Yet these diverse discourses also "constitute contested frontiers, inasmuch as they exist by virtue of the boundary" (Newman and Paasi 1998, 196). As the editors of this volume suggest, discourse crosses borders – just like people do.

As depicted in much Canadian and American nationally canonized literatures, indigenous Americans have been feared for their ability to enact violence on the body as well as the spirit or mind. This fear has been part of a long-running paranoia.[3] Jodi Byrd (2011, 207) notes that from the colonial era to the present, "the United States is haunted by the spectres of its origins, and the displaced narratives that have been continually rewritten do not altogether disappear. America becomes obsessed with borders and its frontiers." As a result, borderlands' native inhabitants were thought to abduct civilians by mere seduction. They could strip from a person the "garments of civilization" (Turner 1920, 4), or convert perfectly good white settlers into "native American Calibans" (Marx 2000, 111). These constitute that Eurowestern myth of the borderland as an oversexed wilderness, a refuge for the illiterate and incontinent savage that defies the rationale of literacy. Geographer David Sibley (1995, 183) argues that borderlands' significance stems from their status as both "zones of uncertainty and security." In the colonial imaginary, natives personify transgression – that is, the crossing of political, cultural, racial, and moral lines into a region of danger.

As such motifs endure, the borderland remains a problem of discourse; but combating misrepresentational stasis means imagining new narratives. Current borderlands literatures are rife with possibility partly because their authors riff on frontier-territory motifs. Native writers, especially, "continue to resist [an] ideology of containment and to insist upon the freedom to reimagine themselves within a fluid, always shifting [border] space" (Owens 1998, 27). Despite alarming political power, the invented Indian can be re-storied, escaping colonial containment into a margin of misbehaviour and mutability, where cardboard representations are transgressed and deconstructed.

To this end, writers of the northern Plains have reimagined the Canada-US borderland as a site of sacred trespass that transforms body, soul, and language. In their fiction about life near the international boundary, Anishinaabe authors Vizenor (White Earth) and Louise Erdrich (Turtle Mountain), as well as Blackfeet-Gros Ventre author James Welch, turn transgression against the modern national narratives where it is fabricated. Canadian Cherokee novelist Thomas King does the same in his writing about borderland Blackfoot communities. These authors take up two loci of imagined native transgression: the body, as it trespasses against Eurowestern cultural and religious mores (namely by miscegenation and transvestism); and language, as oral forms threaten the dominion of the inscribed historical, legal, and theological word. These writers interrogate the fantasy of indigenous transgression by crafting characters that enact tricky, interstitial personalities, which expose the binate structure of borders – you are either *in* or *out* – as a parent fantasy. This process, which Vizenor calls "erotic conversion," dismantles colonial territoriality by breaking down its oppositions: citizen-savage, Christian-heathen, man-woman, human-alter-species, and so on. In this paradigm, the most effective tool for social change is mischief. Each author insists that, as writers and readers, we enact this conversion on multiple fronts, sacred for its power to make permeable the discursive boundaries that inhibit mutual understanding between indigenous and Eurowestern peoples.

TRANSGRESSIVE IDENTITIES ON THE MOVE

In her work, Erdrich interprets the Great Northern Railway as a line of Euroamerican empire that bisects Turtle Mountain Chippewa lands in North Dakota and Montana. The narrator of the poem

"Indian Boarding School: The Runaways" imagines this parallel boundary as "old lacerations" or scars across the face (Erdrich 2003, 19) – here Erdrich argues that borders are inscribed on the land and on the bodies of indigenous peoples. Railways, like their ethereal, cadastral counterparts, have carved up the continent with sociospatial demarcations; they figure prominently as divisions in stories about the Canada-US border. Delineated as an obstacle to capitalist manifest destiny, the catch-all "Indian" had to be (re)moved to reservations, out of the way of the transcontinental railroads and out of new territories put to "proper" use by Eurowesterners who viewed the land first and foremost as a "locus of ... economic ... value" (Marx 2000, 150). Colonial cartography inscribed self-legitimizing definitions and categories, and wrote the transgressive indigenous body into being. Reservations work in this very manner: fabricated boundaries immediately make trespassers out of migratory or diasporic peoples. They are suddenly "out of place."

All borders serve nation-states as mechanisms of categorization and separation. Plains historian Elliott West (2004, 9) recognizes that they emerge from "an essential part of state-making," which is "the government's power to catalogue and name its constituent parts and to set standards for citizenship. Partly this involves the government deciding for itself who is in and who is out, politically speaking." States then delineate and dispense "unalienable rights" according to their own systematized definitions. At international boundaries, indigenous people suffer multiple foreign states and thus subsist between imperial rocks and colonial hard places. Natives persist, however, in what Kevin Bruyneel calls the "third space of sovereignty," a space created by colonialism that always refers back to and indicts colonial power. Discursively, the third space "resides neither simply inside or outside the American political system but rather exists on these very boundaries" (Bruyneel 2007, xvii). Borderlands, then, are a contiguous "out there." They are political and legal and economic limbo.

This limbo has not gone without consequence for native self-conception, either. Perhaps the most worrisome outcome of (neo)-colonial cartography is that its inscription *leaves scars*. West (2004, 11) observes that in both Canadian and US contexts, "the state's imposition of categories on native communities ... indirectly precipitated basic changes in how native peoples thought of themselves." These categories cast natives on international borders as transgressive in (at least) two different narrative frames. Further, the process by which national discourse constructs and imposes identity "is itself political action and is part of the distribution of social power in society" (Newman and Paasi 1998, 195). This only enables other violence. Thanks to the generalization of the "Indian," when reservation lines were drawn (whether at the behest of corporations or national governments), indigenous groups were corralled onto reserves without any regard for the relations between them. Longstanding conflict between the Blackfeet, Cree, Gros Ventre, and Assiniboine pervade the disquieting atmosphere of Welch's *Winter in the Blood* (1974), for example. In 1888, the United States enclosed two bitter rivals, the Gros Ventre (or A'ani) and Assiniboine, on the Fort Belknap Indian Reservation, where the novel is set. Such moves result from and reflect the circumscription of diverse peoples into a singular category.

Cadastral scars are evidence of remapped social, political, and even physical space. The anonymous narrator of *Winter in the Blood* reflects on the metamorphosis of his family's land, and locates the change in a fence, one of many markers of Eurowestern "progress": "The fence hadn't been here in the beginning ... But the ... cottonwoods and willows, the open spaces of the valley, the hills to the south, the Little Rockies, had all been here then" (Welch 2008, 126). He regards the fence as the first thing that does not belong, a mark of individual land tenure that has divided his community just as it has divided the land. Like all borders, fences are "contested cultural and symbolic manifestations of territoriality" (Newman and Paasi 1998, 188); they have a major impact "on the formation of sociospatial identities" (201). Other scars, like locomotive rails and names written into wet concrete, similarly remind Erdrich's runaways of the compound political "punishments" and "delicate old injuries" their people have endured (Erdrich 2003, 19). When a people has been written into national narratives, those "most widely circulated versions of the past," as antithesis and even obstacle to civilization, they suffer deep psychocultural injury (McCormack 2005, 111).

Although in some cases they lie within two national spaces, native people at the Canada-US boundary are *not-citizens* according to both national stories. Eurowestern polities on this continent have historically recognized only colonial agents, and have converted natives into "fugitive objects" in their own homelands, runaways in a picturesque landscape (Vizenor 1998, 33). Part of drawing the

Indian involves abstracting her, making her generic and unreal so as to project her into the past. This process keeps native people at a distance – in a narrative sense, even from themselves – and on the horizon.[4] Vizenor (1998, 178) objects to "territorial" thinking on similar grounds, not just for its inherent compulsion towards possession but for its engendering of a border horizon where indigenous people become dehumanized. The extension of borders vis-à-vis manifest destiny may be understood, too, as a process by which indigenous nations once externally colonized are narratively subsumed into the colonial system, which thereby evaporates their sovereignty into the abstract space of the borderland (Byrd 2011, 202). Borders anchor a conceptual space ripe for symbolic cultivation where images of the other may be constructed, imposed, and naturalized at will over a given land area.

This is where Bruyneel's "third space" opens. The third space is a sovereign rupture between, but also within, Eurowestern national spaces. It is a place where native people have turned colonial abstractions to their advantage, often through movement. Along the Canada-US border, where foreign narratives and bureaucracies have construed them asymmetrically, moving between national spaces becomes a strategy of resistance by evasion. Sitting Bull deliberately crossed after the Battle of Little Bighorn in order to disrupt military efforts to confine the Lakota to reservations, and so exploited one artificial boundary to temporarily defeat another (West 2004, 12). This was a narrative tactic as well as a politicospatial one, just as Sitting Bull would later dispute the already growing American legend of "General" George Armstrong Custer by re-telling the events of the officer's death. In *Winter in the Blood*, the narrator's grandmother tells a similar story about Heavy Runner leading a band of Blackfeet into Canada to avoid the US Army. As she speaks, "her eyes were [no longer] flat and filmy; they were black like a spider's belly and the small black hands drew triumphant pictures in the air" (Welch 2008, 29). As they enact strategies of narrative resistance, natives re-inscribe themselves as anything but objects of the colonial will – they become fugitives *from*, not *within*, colonial delineation. Historians note how malleable the forty-ninth parallel has been for indigenous people, who first viewed it as a "quiet and unexplained guest ... with its seemingly arbitrary straight line, slightly mysterious origin, and hazy significance" (LaDow 2004, 65). Much as oral tales may transform in the retelling, the "story" of an imaginary border may also be transformed through such a paradigm. The international boundary was once drawn on the land by colonial surveyors and mapmakers – it can be redrawn, and even undrawn, by authors and other cultural workers.

This process involves a kind of indigenous narrative cartography. Borders *can* have good medicine, especially when disarticulated from their Eurowestern sociopolitical contexts. As if in response to Erdrich's poem about runaways on the rails, Vizenor writes in *Hotline Healers* (1997) of Gesture Browne, an unaccredited Anishinaabe dentist who acquires the private railroad of a rich banker near the international border in Minnesota. Like Erdrich, Vizenor alludes to the Great Northern Railway, the only privately funded transcontinental railroad in US history, built by Canadian-American James J. Hill – the "Empire Builder." Now part of the Burlington Northern Santa Fe (BNSF) Railway, it runs from Seattle to Saint Paul on the south side of the border – a region known as the Hi-Line – across many native homelands and reservations, including Fort Belknap. In indigenous hands, the Great Northern becomes the "Naanabozho Express," a "tricky express" train service that runs "at no cost to native passengers" (Vizenor 1997, 149). (Naanabozho is the quintessential border-crosser in Anishinaabe cosmology.) Welch, however, questions such playful re-conceptions: the narrator of *Winter in the Blood* seems at the Great Northern's mercy as he initiates a string of self-destructive sexual encounters in Montana rail towns. He aimlessly wanders the Hi-Line, yet he finally repositions himself with respect to it by reawakening a defiant family history rooted in a particular native geographic understanding.

These stories express an indigenous "migratory identity" (Anderson 1999, 21) – by this, scholars mean both a literal and literary migration, a motive reassertion against colonial narratives. In the mind of the anonymous narrator of *Hotline Healers*, natives are "natural roamers" (Vizenor 1997, 5); Teresa, the mother of Welch's narrator, sees them as "wanderers" (2008, 16). They may *move* differently. Hybrid states, such as mixedbloodedness, reflect this migratory identity by refusing to stay put within established conceptual oppositions. Vizenor (1998, 181) calls this state of flux *transmotion*, a "natural right of motion" that stresses dialogic movement over the static division, possession, and consumption of land. It is not, however, incompatible with an investment in distinct locales, despite how stereotypes of the nomadic Indian have been used to undermine native land rights. Cherokee scholar Daniel Heath Justice (2005, 49) argues that indigenousness "doesn't always require an eternal presence in a particular location: though not necessarily elastic, the relational principle of peoplehood is adaptable to multiple spirits and sacred landscapes. Again, the emphasis is on ... [the] homeward relationships of rootedness and movement." Rather, transmotion involves a conversion of narrative positions, from object to subject, narrated to narrator. It is a form of cultural resistance that articulates indigenous "nationhood, or the idea of a shared culture requiring protection" (Lyons 2010, 119), as it moves.

In North America, tribes' autonomy has paradoxically been dependent on and prescribed by Eurowestern nation-states (West 2004, 10). Sociologist Duane Champagne (2005, 5) of the Turtle Mountain Band of Chippewa argues instead that native identity and nationality "derive from their occupation of the land and from their self-government," not from "colonial legal proclamations." Because native domain derives from residence and a sense of reciprocity with the earth, not tenure or profit-use, it has always cooperated with – not imposed itself on – North American geography. This has not been the case for Eurowestern polities, and so native cultures have always run counter to the border logics of nation-states (West 2004, 9). Indigenous understandings of political and cultural sovereignty likewise run perpendicular to Eurowestern nations and generate conflict at every intersection. Bruyneel's third space still manages to open a space for tribes, "function[ing] as an anti-statist, postcolonial supplement that disrupts the logics of colonial rule" (Byrd 2011, 188). Yet sovereignty is itself a problematic construct shaped by Eurowestern understandings of land, space, social relations, and governance. As a result, the term *"native sovereignty* bears little resemblance to the principles of Native American societies" (Pulitano 2003, 70). The failure of treaties to protect the rights of indigenous people, even against the active abuses of national governments, hinges on these conflicts.

Others have envisioned a conversion of the sovereignty concept. Bruyneel (Byrd 2011, 188–9) claims that the third space "emerges out of indigenous resistances and articulations of alternative governance strategies that remain in spite of the legal machinations of the occupying colonial nation-state." Like Champagne, however, Vizenor (1998, 16–17) sees an antecedent native sovereignty that may exist independent of the colonial: for him, "native sovereignty" denotes transmotion, "the rights of [which] are personal, totemic, and reciprocal; not ... possessory." This conversion is unalienable, and not erotic in the sense of sexual penetration or possession (as was the conversion of indigenous to "Indian") but in the sense of representational flux. Vizenor shows how a play with images, a kind of striptease, can challenge misrepresentation by exposing the "Indian" as mere colonial fantasy. In *Hotline Healers*, mixedblood protagonist Almost Browne develops a new ceremonial *debwe* dance to triumph over such gross cases of mistaken identity. His *oshkidebwe* ("new truth"), now an ironic masquerade of masks, costumes, manners, and poses, undermines Eurowestern classification. Almost does the same work when he removes his sweater during a radio interview to reveal a black lace bra.

Osage scholar Robert Warrior (1995, 123) describes native sovereignty as a praxis and a "decision we make in our minds, in our hearts, and in our bodies." It is a process of enacting a conviction of self-determination, but it must work in a reciprocal mode, not an imperial one. Byrd (2011, xvi) shows how native self-determination can be both independent from and inflected by Eurowestern colonialism, and she explains that indigenous sovereignty is "found in diplomacy and disagreement, through relation, kinship, and intimacy. And in an act of interpretation." Native sovereignty exists, in part, through the sharing of stories. The trademark "diplomacy" of colonial treaty negotiations is a violent misreading of reciprocity, yet even this may be converted into another layer of retelling.

Vizenor, Erdrich, King, and Welch lend a sharp eye toward the relationship between personal and community sovereignty, toward the ways that story helps construe the native body politically. Like *Hotline Healers*, Erdrich's *The Plague of Doves* (2008) is framed by a case of mistaken identity: when three indigenous men and one boy are believed to be guilty of the murder of a rural North Dakota family, a band of white vigilantes captures and lynches them. The vigilantes' conviction derives from an unwavering confidence in the authenticity of a national fantasy, America's savage "Indian." By hanging them, the local men symbolically possess and erase their indigenous neighbours, writing them back into their Eurowestern cultural narrative as necessary casualties of civilization. Escape from narrative enclosure depends on the lone survivor, Seraph Milk, finally telling his story. Discursive borderlands like these are an "aesthetic earth out of balance" where misrepresentation is disrupted, whether by Almost's ironic identity play or Seraph's sober re-narration (Vizenor 1998, 29). Always palpable in the erotic conversion is that shifting space where cultural simulacra have been laid bare.

EROTIC TRANSFORMATIONS OF BODY, LAND, AND COMMUNITY

Erdrich's characters struggle with the complex politics of mixed communities and recognize how easily an "unsatisfactory present" can "become [an] entire history" if no action is taken. Their battles reflect those of real native people deeply invested in their border communities and homelands. Their triumphs and (often mortal) failures demonstrate the urgency of political action and how a concern for storytelling must be a part of this action. As if trying to speak such change into being, and to dissolve the bigotry that isolates him from his Anglo girlfriend in *The Plague of Doves*, mixedblood narrator Antone Bazil Coutts reveals the epitaph he has even at a young age written for his headstone: *"The universe is transformation"* (Erdrich 2008, 282).

Indeed, transformation is a fundamental part of indigenous life, self-government, and literary discourse. Champagne (2005, 7) notes that "native institutional relations are not seen as unique and separate from the cosmic order of life and change, but are ... embedded within the cosmic order, and reflective of cosmic order and direction." Change is, perhaps ironically, vital to the sustainment of natural order. Native borders are therefore not colonial, because they proceed from nature and defy territorial logic by accounting for natural change and movement. In his autobiographical essay "Crows Written on the Poplars," Vizenor (2005, 105) meditates on the tenuous border that separates him from a common red squirrel he has killed: as a mixedblood, Vizenor too has "been the hunted, to be sure, cornered in wild [colonial] dreams, and he has pretended to be a hunter in his stories, but he has never lived *from the hunt*" (emphasis mine). Here Vizenor opens up a space for erotic conversion by inhabiting the conceptual barrier or borderland between hunter and hunted, self and other. "The hunt" is an active space that only exists in their relation.

Two-spirit criticism serves as a useful tool for unpacking conversions whose locus is the indigenous body. Cherokee poet Qwo-Li

Driskill (2010, 73) relies on the logical openness of two-spirit criticism in his analysis of heteronormativity; in this, he explores "the ways colonial projects continually police sexual and gender lines" – borders and territorial logic are at work here, too. One realm of enforcement is discourse, which includes literary narrative and the institutions that canonize particular literatures over others and thereby fabricate native absence. Eurowestern thought views those with two identities (e.g., genders, ethnicities, nationalities, spiritualities) as aberrations: not belonging to *both* categories, but to *neither*. Conversely, the term *two-spirit – niizh manitoag* in Anishinaabemowin – communicates "numerous tribal traditions and social categories of gender outside dominant European binaries" (Driskill 2010, 72). It relies not on mutual exclusion but on the syncretic spirit of bothness. Two-spirit criticism works towards decolonization by laying bare colonial practices and borders of all kinds.

Conversions of the body occur throughout *Truth and Bright Water*, *Hotline Healers*, and *The Plague of Doves*, as well as Erdrich's *The Last Report on the Miracles at Little No Horse*. Characters cross-dress and miscegenate, including with other species. Native women develop questionably intimate relationships with Catholic priests. When the erotic conversion manifests in sex, it does so in that point of convergence between sex and religious prescription. In part, this is a literary response to preconceived native transgression by Christian missions in the Great Lakes and Plains regions. Father Jean de Brébeuf wrote in seventeenth-century New France that there was "one thing ... which might give apprehension to a Son of the Society, to see himself in the midst of a brutal and sensual People, whose example might tarnish the luster of the most and the least delicate of all the virtues, unless especial care be taken – I mean Chastity" (quoted in Blackburn 2000, 60). In *The Plague of Doves*, a bishop issues a similar warning to his diocese about Métis women: he advises priests to "pray hard in [their] presence ... and to remember that although their forms [are] inordinately fair their hearts [are] savage and permeable" (Erdrich 2008, 11–12). Brébeuf and other Jesuits helped reproduce this perverse and pervasive belief – that sex among native people is sinful and warrants shame – in Eurowestern discourse. A nameless priest in Harlem, Montana, one of Brébeuf's fictional analogues in *Winter in the Blood*, exhibits this irrational fear in how he attends to the Blackfeet. He will not step foot on the reservation, even to administer last rites and bury deceased Catholic natives on their own land, for fear of being seduced by a way of life that he regards as an inherent proclivity for moral transgression (Welch 2008, 47). For the priest, the reservation border keeps things *in*, not out.

All four authors appear intent on provoking these artificial lines, but in subtly different ways. Thomas King engages religion directly on the basis of representation in *Truth and Bright Water* and casts as his tricky figure the "famous Indian artist" Monroe Swimmer, who catalyzes the novel's mystery plot by appearing to two young boys as a woman swan-diving into the Shield river. It turns out that Swimmer – his name a pun and reference to the famed Cherokee medicine man (Conley 2005, 38), eponymous source of James Mooney's *The Swimmer Manuscript: Cherokee Sacred Formulas and Medicinal Prescriptions* (1932) – habitually wears a braided black wig, one of many plays on cultural stereotype in the novel. But the primary reason for Swimmer's return to his home reserve after years on the Indian art circuit is to rub out "Christian mission presence ... by painting the old church in trompe l'oeil clouds and sky so that it is almost indistinguishable from its natural surroundings" (Weaver 2001, 87–8). The illusion becomes a way for his people to imagine themselves outside the shadow of moral and sexual regulation.

Almost Browne is born from the union a native nun and a white priest, the remainder of an ironic equation that is part of a process of imagining possibility. Almost hosts his *oshkidebwe* heart dance at universities, which are "some of the best ... sources of tricky stories" (Vizenor 1997, 52). But the narrator, Almost's cousin, says that the *debwe* began with an Anishinaabe woman in the fifteenth century who engendered the first "erotic conversion of monks and animals, a pleasurable passion at the first monastery" to restore natural harmony on the Plains (Vizenor 1997, 6). The Jesuits detail their collective sex acts with other species in the fictional *Manabozho Curiosa*, a rare monastic script full of sensual ceremonies – what the narrator calls the "very first native manuscript," reorienting even what it means to be "native" (Vizenor 1997, 148). These passions are conversions of body and spirit from institutionalized Christianity to something hybrid and new. After their experience at the headwaters of *gichiziibi*, the "great river," the monks establish a new faith free of Catholic prescription. Their former, nameless abbot dies naked and mysteriously frozen into the river, a broad blue smile on his face – but the monks discover he was secretly a woman all along.

They consecrate him Saint Hilaria, after one of many cross-dressing saints in Catholic tradition, who lived much of her life as a eunuch until she died and her identity was discovered. The monks name Gaaskanazo, the old woman who hears the stories of other species, for their new abbot. The monks were converted, stripped of their colonial garments, and could not go back to the Society, for "the braces of monotheism [had been] overset in erotic trickster stories" (Vizenor 1997, 165). Almost, too, begins to cross-dress as a nun as part of erotic acts with other species.

The Last Report on the Miracles at Little No Horse (2001) hinges on a written confession to the Pope by Father Damien Modeste, who has served the Ojibwe at Little No Horse for more than eighty years – as a woman. Father Damien (formerly Agnes DeWitt, and well beyond one hundred years old by the time of his writing) intends to drown himself at the bottom of a lake; but before he can carry out his plan, he laughs himself to death on the shore. It is only by an old friend, Mary Kashpaw, who has followed him, that the process of his watery burial is completed. It was Agnes' near-drowning in a Red River flood, also, that precipitated her conversion to Damien in the first place. Rivers, lakes, rain, and other natural fluids like semen serve as a medium for conversion, from states of containment to states of bothness. When a life-or-death footrace must be run to settle a dispute between two enemy groups in the novel, an "ikwe-inini, a woman-man called a *winkte* by the Bwaanag [Dakota]" removes his deer-hide dress to reveal "nothing but a white woman's lace-trimmed pantalets" (Erdrich 2001, 153–4). Though they show some contention over whether the *ikwe-inini* qualifies as male or female for the purposes of the race, the assembled group agrees that the boy may run.

Erdrich's indigenous characters understand these phenomena as part of a new "Métis Catholicism," whose hypothetical practitioners achieve spiritual growth by questioning religious law. Religious traditions and mores are sacred but not static; they are transformable, and more sacred in the process of transforming. In *The Plague of Doves*, elderly brothers Mooshum and Shamengwa "liked to speculate about the form that Metis Catholicism would have taken and whether they might have had their own priests. Mooshum insisted it would be better if the schismatic priests were allowed to marry, and Shamengwa was of the opinion that even Metis priests should keep their chastity" (Erdrich 2008, 30). The novel begins when Evelina's great-uncle, "one of the first Catholic priests of aboriginal blood," develops a blended religious ceremony in order to intercede between a mixed community and the eponymous "plague" which has descended upon them. By chance a dove wounds Seraph Milk (a.k.a. "Mooshum") during the ceremony, but this leads him to his future wife. Native residents of fictional Pluto, North Dakota, believe that finding a lover is in fact the "best reason" to go to church – or so Seraph advises his granddaughter. Evelina duly enjoys her first masturbatory experience while she daydreams about Corwin Peace, a boy from church, but she later leaves the oppressive small-town atmosphere to attend the University of North Dakota, where she interacts with border-crossing poets and mixedbloods, and falls in love with the erotic diaries of her new French muse, Anaïs Nin. The university, as a kind of nexus of narrative diversity – among these, the perturbing phenotypic differences between "AIM-looking" and "middle-class BIA Indians" like Evelina – prepares her for later transformation. Monotheistic institutions tend to quash diversity and re-inscribe all into an "us versus them" opposition.

Hierarchical "monologues" like Catholicism cannot undergird conversions of colonial representation, for they are its creators. Evelina suffers mental illness engendered by the schism of an obsessive relationship with a fellow female patient and transvestite at the psychiatric hospital, Nonette. Like her previous experiences, this one is monologic in its obsession; though her desires are unauthorized, their possessive, univocal spirit precludes them from efficacy. The experience does, however, precipitate Evelina's healing reunion with her home and a farewell to Nin's erotic diaries. Evelina must leave her desires behind, for possession enforces a single voice and conversion may take place only in erotic conversation. Similarly, the hegemonic voice of the Great Northern Railway drowns out those of the mixed rail towns and reservation communities along its route until characters reimagine their relationships to the borderland.

The voice of home is truly that of the border *land*, a material fulcrum around which native inhabitants and Eurowestern squatters battle each other, each fighting for a different notion of peoplehood. The stories discussed here all direct native people to rely on the earth in this conflict, because its existence precedes the abstractions that create borders and border towns. Erdrich's Pluto is vexed by the fact that its Euroamerican founders stole tribal land through suspicious surveying – the same may be said of Havre and the other Hi-Line towns in *Winter in the Blood*. When its narrator buries his grandmother, the family matriarch and story keeper, he plants an oral history in the soil, a narrative medicine that encompasses and shapes him and therefore roots his very being in the earth, defying dispossession. He learns that body and word share their materiality with the land, and that their record may be preserved and accessed through careful excavation – not by anthropologists, but by storytellers.

Dwelling *on*, *in*, and *with* the land is an essential part of transmotion. Much the way indigenous crop rotation practices kept American soil healthy for centuries, migration revitalizes the land and peoples' relation to it, and keeps them from stagnating. Migration stories also have this benefit: by definition, they speak of *home* as much as they speak of travel (Lyons 2010, 4). In *Truth and Bright Water*, Tecumseh's father Elvin is shamed by his community because he has agreed to smuggle bio-hazardous waste from Indian Health Service (IHS) hospitals across the international border, where he dumps it on the reserve. Though his smuggling might otherwise serve the decolonization imperative, respect for the land in which one dwells takes precedence here. It is no coincidence that the waste brought to Bright Water is refuse from a federal agency and sick indigenous bodies.

Movement and illness are deeply connected through the body. Running indicates good health, and it functions as a healing practice in many indigenous communities. This is reflected most famously in N. Scott Momaday's *House Made of Dawn* (1968), but it also appears in the Indian Days footrace in *Truth and Bright Water*. Lum spends much of his time practising for the event in order to beat "that Cree guy from Hobbema [in Alberta]" (King 1999, 227); the Cree became enemies of the Blackfoot Confederacy in the nineteenth century, and the race thus symbolizes and reinforces communal bonds. But Lum will not race at Indian Days, because his father Franklin wounds him and gives him a bad leg, one of many signs of domestic abuse in the novel. For King, body dysfunction reflects family and community discord, as well as a personal mental distress that encourages Lum's aggression and angst in the novel. Yet movement itself may also indicate psychological illness. It may be a response to "[Eurowestern] nationalist imaginings," which work as a representational constraint that fractures indigenous identity into private and public cultural, legal, and political fragments (Christie 2009, 178). The narrator of *Winter in the Blood*, his namelessness denoting identity crisis, travels east-west along US Highway 2, aimlessly bedding women so as not to feel his disaffection. His movement is motivated by the theft of his gun and electric razor by his Cree girlfriend, whom he is searching for, and it echoes the mixedblood drifter that he believes to be his grandfather. Both symbolize that characteristic "nagging sense of 'distance'" of disunited Plains border country (Teuton 2001, 626), what the narrator describes as an "event of distance." His movement offers him no healing, for the women are dislocated erotic generalities, representative only of his isolation and not the voices of their home spaces – Havre, Harlem, Malta. The encounters instead correlate unbalanced sexual relationships with an unmoored identity. Only a change in the way he understands human relations will help the narrator traverse his alienation and anchor him in story and place (Welch 2008, 127). Balanced, reciprocal movement (or transmotion) is a conversion from separation and dysfunction to unity and well-being.

These characters feel especially "distanced" when they have yet to awaken to the borders that limit and fragment their lives. Transgression may in fact open up new experience and awareness beyond the bounds of colonial control. Evelina Harp is overwhelmed during a three-day acid trip by serpents that crawl out of the electrical sockets, for instance, a familiar Christian metaphor for decadence that demarcates sinner from saint. These images work alongside the erotic presence of Nin and together resonate with Jesuit prescriptions about sexual behaviour among "exotic" native women on the northern Plains. Beset by this representational violence, Evelina becomes "unaware of who [she is]" – it is only on day three when "a comforting, old friend" shows up, the eastern tiger salamander, that she starts "to sense a reliable connection between one moment and the next, and to feel with some security that [she] inhabited one body and one consciousness" (Erdrich 2008, 225). Evelina's fragmentation denotes ill health. When earlier as a child she traces the name of her pubescent lover across the secret parts of her body, striving for a million repetitions, her mantra culminates in an inscriptive, masturbatory transgression as she orgasms in the bathtub, submerged in water like Saint Hilaria. But Evelina's "alphabetic orgasms" lack self-awareness; she cannot recognize the borders she has crossed. Until the serpents and her intimate encounter with Nonette in the hospital drainage tunnel, she will not understand the border-troubling transformation she has begun.

In the worlds of these Plains authors, an awareness of transgression does not elicit shame. The Jesuits' awareness of their cooperative trespass, a miscegenation of Christianity and nature, marks a cataclysmic disavowal of borders in both the shared acts and the monks' later inscription of the event. This is no calm conversion but one rife with orgasm, and thus extreme unrestraint and risk. The risk is embodied and narrative, for Christianity has been wielded against natives as another absolute doctrine that "lock[s] the true believer into a moral system that lacks imaginative freedom" (Ruppert 1995, 94). Such doctrines set people into fixed beliefs that impose unchanging imaginings of the other, locking natives into stories that also refuse change. Only narrative re-imagination opens up a space for liberty.

THE XENOGAMOUS TRESPASS

Almost Browne is of mixed ancestry, and his birth in the backseat of a station wagon on the side of the road at the edge of the White Earth Reservation – "almost" on the rez – reflects his personification of the borderland. His father is believed to be an unnamed reservation priest, while his mother is a reformed Anishinaabe nun named Eternal Flame Browne – Almost is "almost brown," a racial ambiguity in Eurowestern frameworks (Vizenor 1997, 11). Perceived as a kind of identity crisis, mixedbloodedness may instead serve indigenous authors as a mutable subject position for the unsettlement of colonial categories. It raises miscegenation as an issue of religious, cultural, and political transgression, interpreted asymmetrically on either side of an international boundary. *The Plague of Doves* centres on the Turtle Mountain Chippewa, many of whom identify as prairie Métis ("mixed blood"). As a people of the Red River basin in Manitoba, North Dakota, and Minnesota, Métis people have different standing in US and Canadian contexts. The Métis have legal status in Canada but do not in the United States. Layers of ambiguity present in the designation "Métis" work here, like other language of mixedbloodedness, as a way to both interpret and *almost* deconstruct narrative misrepresentation.

While Vizenor's circuitous wordplay differs from the bleak, barebones quips of Welch's narrator, each foregrounds linguistic ambiguity. Erdrich tends to take Vizenor's tack: in *The Plague of Doves*, a boy joins his Ojibwe grandfather Mooshum in teasing Father

Cassidy, who while visiting the family cites Latin etymology to explain "concupiscence," his arrogant label for their sins. The priest warns against "any act of imaginary or ejaculatory fornication," but Joseph, the boy, turns Cassidy's discourse back on the Church to tease its xenophobic anxieties. He deploys a joke etymology: fornication, "from the Latin *forn*, as in *foreign*, for relations with foreigners" (Erdrich 2008, 26). Mooshum and Shamengwa then toast Joseph with whiskey, to the dismay of the priest. Wordplay works in resistance to institutional discourse, one tool Catholic missions and Eurowestern governments have used to limit indigenous rights. Etymology, which validates language via genealogy, artificially separates natives from literacy.

Miscegenation laws are rooted in an anxiety about the artificial border that defines the "savage." The Church's warnings about the seductive powers of native women belie, ironically, a colonial fantasy of miscegenetic penetration of the new "virgin land" and the indigenous other (McClintock 1995, 24, 30). Native women provide the perfect point of contact with priests for authors of the Plains and Great Lakes, regions with a long history of Catholic missions. As they play out colonial anxieties in their texts, they expose the sources of fraudulent representation. Almost's white priest father and native nun mother work this way; so do the Jesuit monks' masturbation with other species. In *Winter in the Blood*, the narrator's mother, Teresa, develops a curiously intimate relationship with the priest in Harlem as they exchange mysterious letters and then spend time alone at his home. Her son suspects them of having an affair, and he construes the relationship as a betrayal rather than a strategy of self-determination or resistance (Welch 2008, 47, 58). His suspicion reflects an internalization of colonial fears and desires about the indigenous body, as well as colonially inflected tropes of indigenous betrayal.[5] The reservation community's suspicions about Almost's nameless father in *Hotline Healers* work the same way.

Xenophobia, over which the erotic conversion triumphs, relies on this very paradox of exoticism. For both the xenophobic and xenophilic impulses, exotic representation functions as a "barrier ... of a petrified kind which transports the [native] body into the world of legend or romance" (Barthes 1972, 84). Exoticism is a masquerade of stereotype that functions to disguise and therefore erase, and this is a kind of sterilized eroticism that "negates the flesh ... [like] a vaccine" (84). The native body thus disappears behind the "Indian."

King teases this effect by revealing the Indian from within an icon of white American beauty. Lucy Rabbit, a Blackfoot woman in *Truth and Bright Water*, believes Marilyn Monroe "was really Indian and that she was adopted out when she was a baby" (King 1999, 19). Lucy is anxious to look like Monroe, yet argues that she wants to bleach her hair because "Marilyn was ashamed of being Indian ... that's why she bleached her hair ... so [she] can see that bleaching your hair doesn't change a thing" (201). Lucy wants to break down the border of phenotype, but she only manages to dye her hair "flaming orange" over and over again. Through this, Jesse Rae Archibald-Barber (2009, 244) argues that Lucy also "crystallizes the image of the colonized Indian maiden as a Hollywood idol." In Cherokee tradition, Rabbit is a tricky figure who survives by failure and chance, not cunning; Lucy's failed trick places the exotic other *inside* the white female icon. Exotic representation, and the abstract horizon it populates, must be teased. It must be stripped of colonial fantasy, as in *Winter in the Blood*, where all romance collapses to make way for local histories. The native body must travel a local discursive borderland, not the exotic horizons of the Eurowestern imagination.

Miscegentic anxieties manifest in animal representation because the separation of human and nonhuman animal, more staunchly defended than even the border between European and "Indian," forms one cornerstone of Eurowestern civilization. Images of other species have long been deployed in scientific, religious, political, and popular discourses as a comparative tool for racial domination; they make up the metaphoric palette by which colonial narratives paint transgression on others. This also reduces nonhuman animals to figurative weapons in human language battles, an erasure of their sovereignty as living beings. The term *xenogamy*, rather than miscegenation, better covers the range of transgressions that looms on the colonial narrative horizon, a threat-fantasy of complete union with the nonhuman other – or *bothness*.

One common animal metaphor used against native North Americans is the snake, which in Western Judeo-Christian iconography symbolizes a transgression singularly sexual, devious, and prideful in its defiance of social hierarchy.[6] As a child, Evelina is inculcated with an ideological paradigm that imagines indigenous persons as snakes. She rattles off a sequence of textbook axioms that satirize rationalist classification during a presentation to her Catholic school class, for instance: "Snakes live in holes. Snakes are reptiles. These are Science Facts ... Snakes are not wet. Some snakes lay eggs. Some have live young" (Erdrich 2008, 48). This is the source material for her later cobra acid trip. Sister Mary Anita presses Evelina for information on local species like the Plains garter snake, as if to stress their indigenousness to the children. She seems pleased, too, with Evelina's recitation of their Latinate species names, as the Chippewa girl reproduces the linguistic tools of her oppression.[7] The nun plays her part in an iconographic duel: she has caught and punished Evelina for drawing a "Sister Godzilla" cartoon during class, a silly visual metaphor that teases the border between Judeo-Christian and native iconographies and even pop culture (Erdrich 2008, 45, 48–49).

Despite its distaste for reptiles, in Erdrich's world the Church reviles amphibians even more. Part reptile, part fish, and endemic to the prairie borderlands, the black salamander is emblematic of a border existence. These "lumps of earth" symbolize a constant trespass against the Church: the nuns "believed [salamanders] were emissaries from the unholy dead, sent up by the devil, and hell was full of them" (Erdrich 2008, 29). Yet they share a kind of kinship with the Harp family. Evelina and Joseph enlist their father to help dig a backyard pond for the black amphibians, which they collect from the puddles at school; one of these "comforting old friends" later saves Evelina from her reptile hallucination. The salamander becomes a liberating image for the Métis because it thrives only by existing in both land and water environments. In *Truth and Bright Water*, a lost duck fills a similar conceptual role, as a transported Southeastern earth-diver myth figure: dwellers in two worlds, the earth-divers must travel the depth of a primordial water to bring up soil to form land for humans. This watery barrier functions like the river that divides the novel's American and Canadian communities.

Indigenous writers recognize the importance of other species in deconstructing colonial borders equally imposed against nature (Vizenor 1998, 142). Evelina's figurative dialogue with "Sister Godzilla" betrays a secret crush that presages her epiphany at the psychiatric hospital. Another character in Erdrich's ensemble novel, Marn Wolde, engages in erotic acts with the snakes she handles for her abusive husband's Pentecostal-style church. These acts precipitate her use of rattlesnake poison to murder him, which allows her to escape a physical and psychological abuse that escalates in direct relation to the religious influence her husband wields in the community. Not only do Marn's snakes serve as accomplices to her escape, they aid in the emotional transformation that allows her to imagine breaking free of her dysfunctional marriage. Animal representation is a vehicle by which Erdrich's women, like the monks in *Hotline Healers*, overturn the confinements of shame Judeo-Christian religion has instilled.

The nonhuman animal *voice* plays a fundamental role in conversion, too, as do the voices of native people and the earth. In *Winter in the Blood*, the narrator's grandmother and other characters who embody tribal history demonstrate that they can converse with the natural environment. Yellow Calf, an old blind man who reveals himself to be the narrator's lost grandfather, speaks with deer and other animals. His grandson's alienation stems partly from a lack of this connection: the narrator dwells on an advertisement for a fishing lure that calls to fish "in their own language," but he does not believe fish still live in the Milk, the river that curls along the Hi-Line (Welch 2008, 9–10). Here Welch uses a subtle joke to tease Eurowestern assumptions about such relations between native people and other species. Meanwhile, over Panda Radio airwaves Vizenor's eponymous hotline healers are cell phone-equipped native shamans on the reservation, always in motion, always ready to impersonate a bear or beaver for a caller in Minneapolis seeking communion with animals (Vizenor 1997, 120). All of these natural stories encourage conversion against colonial thinking, histories, and even genealogies. As she recalls how so many crows alighted on her station wagon when Almost was born, Eternal Flame assures her son that he was "probably not the only kid to have a junk car as a mother, and ... probably not the first kid to have a crow as a father" (Vizenor 1997, 10). Weaving human and non-human families together dismantles the hierarchies of "civilization."

DIALOGIC TEXTS AND OTHER WORDED TRANSGRESSIONS

In northern Plains literatures, regenerative histories spoken through indigenous voices square off against printed, "official" histories that are mostly static. In *The Plague of Doves*, Pluto's historical society grounds the local community and yet is always at odds with its native residents, who use language games as a chief strategy. Almost and his cousin start Wiigwaas Trade Books in *Hotline Healers*, selling blank books (even fake autographed copies) on the University of

Minnesota campus in order to subvert Eurowestern fetishization of the printed text (Vizenor 1997, 24–5). It is important to recognize that the novels discussed here are books as well as speech acts: all are narrated by characters – *The Plague of Doves* rotates through many – making the link between voice and inscription central to a full understanding. Native texts of the border like *Truth and Bright Water* read "history as story, and story as history"; their authors argue that indigenous people "can own *both* orality and literacy, story and history" (Ridington 2006, 288, 292). This is what makes them so threatening.

Since European contact, illiteracy – that is, restricted access to authorized forms of language and history-writing – has rendered natives transgressive. But historical texts are not always alphabetic or printed. *Truth and Bright Water* is very much about rewriting historical texts that are as varied as naturalist oil paintings or the Rodgers and Hammerstein musical *Oklahoma!* but that nevertheless heavily inform our understanding of the world. Liberation becomes possible when Monroe Swimmer restores Indians in landscape paintings, or when Tecumseh's mother stars in a native adaptation of *Snow White* in which the dwarfs are Indians. These texts suggest an active engagement with history, a continual revision that consults diverse sources in its re-narration of the past.

Native authors recognize, like Evelina Harp, that "when we are young, the words are scattered all around us. As they are assembled by experience, so also are we, sentence by sentence, until the story takes shape" (Erdrich 2008, 268). We embody our personal and community histories, scars and all, and this recognition is part of the erotic conversion's ability to deflate colonial power. When Evelina traces Corwin's name in her own blood, she opens to a process that first "held for [her] the sacred resonance of those Old Testament words written in fire by an invisible hand. Mene, mene, teckel, upharsin" (Erdrich 2008, 10).[8] The erotic gives her true access to the sacred. The bloody inscription that inspires her "alphabetic orgasms" is migratory; though incomplete, it starts the metamorphosis of the native body into a renewed presence.

The *oshkidebwe* is primarily an enactment of sacred words, and the *Manabozho Curiosa* proves that written texts can bear the transformative power of oral ones. A major part of Vizenor's project is to mediate between these two realms of discourse. James Ruppert (1995, 94) observes that "for Vizenor, contradiction is the essence of oral tradition in its emphasis on variation; its play between text and interpretation ... its subversion of absolute definitions of reality, and its ability to guide without demanding" perform a kind of "psychic healing" of the damage done by colonial textuality. Oral stories, in particular, "guide without demanding," much the way the Shield, Milk, Red, or Mississippi rivers do. The *debwe* ceremonies and their *Curiosa* imagine alternate histories of Christian presence on the Plains, which rely not just on oral and written discourse but on two essentially different conceptions of language. Stories of erotic conversions are best crafted among groups of storytellers, like the masturbating Jesuits in the forest; the academic attendees of the final *debwe* at the University of California; or the actors, writers, and technicians involved in the indigenous *Snow White*. These express dialogic as well as syncretic textuality. Susan Friedman (1994, 127) notes that "at the heart of ... postmodern political syncretism is the necessity of keeping [multiple] readings in play," and it is this that, in part, underlies transmotion.

These authors participate in a project of literary nationalism, and recognize that expressions of cultural sovereignty can eke out native spaces in complex political zones such as the forty-ninth parallel. Syncretism does not denote political acquiescence. Jace Weaver and others (2006, xv) problematize even their own language by pointing out a troubling contradiction: "[Nationalism] ... describes a phenomenon that has given rise, on the one hand, to ... the thirst for liberation of oppressed people around the world, and, on the other hand, some of the worst forms of political repression and xenophobia in human history." Yet, inhabiting this contradiction has been fruitful for contemporary indigenous authors. Weaver observes the "explicitly activist" project of native writers that seek to redefine indigenous "nationhood," which differs from what Justice (2006, 8) calls the "coercive nationalism of industrialized nation-states." Native "nationalist" narratives are not monologic – they do not exclude but rather often thrive on irony and dispute – yet they are nationalistic in that, to succeed, they *must be on indigenous terms*. As Stuart Christie (2009, 31) notes, "contemporary indigenous literary expression exceeds the demands of the Anglo-European nation in discourse. The sovereign adumbrations of that Other Nation – cultural, technological, economic – emerge visibly within an *indigenous* sovereign-nationalist horizon of expectations." Native authors work to (de)stabilize the imbalance of discursive borders, particularly near major political boundaries where the lines between indigenous and non-indigenous may seem more complicated and present.

Louise Erdrich, James Welch, Thomas King, and Gerald Vizenor recognize that the old imagined "dangers" of the borderlands have the ability to effect ethical sociopolitical change, if they can only be converted. Narrative automobility, which in indigenous hands revises terminal histories, has always been vital to the health of native self-conception and community identity. Literature that arises from the contradictions of national borders takes aim at territoriality, at national myths and their unethical assumptions, through its expression of a diversity of voices that speak (and dance, fornicate, and cross-dress) from the margin. The "erotic conversion" over the alienation of the Canada-US borderlands must first explore representation and story because, as King says, "the truth about stories is that that's all we are" (2003, 1) – whether native or Eurowestern, American or Canadian.

NOTES

1 Robin Ridington (2006, 291) argues that the name of King's protagonist ironically alludes to both the Shawnee resistance leader and cultural hero Tecumseh, who attempted to assemble a pan-native alliance in the early nineteenth century; and the Civil War general William Tecumseh Sherman, "whose 'march to the sea' devastated former Cherokee homelands in Georgia" and whose name famously represents an appropriation of native culture and history.
2 This conception is evident even in early Puritan American texts and becomes most egregious in popular nineteenth-century US literature. In the work of James Fenimore Cooper (1898), for example, indigenous people on the prairie are "imps" (37, 59, 212, 271), "demons" (35, 58, 266, 383), "specters," and "devils" (34, 61, 62, 136). Such descriptions characterize the "Indian" as not only savage but as an agent of sin and evil.
3 The American captivity narrative genre, inaugurated by *The Narrative of the Captivity and the Restoration of Mrs Mary Rowlandson* (1682), springs from this paranoia of the frontier. Margaret Fuller's *Summer on the Lakes* (1843, 5) provides another example from a female perspective: "For continually upon my mind came, unsought and unwelcome, images, such as never haunted it before, of naked savages stealing behind me with uplifted tomahawks; again and again this illusion recurred, and even after

I had thought it over, and tried to shake it off, I could not help starting and looking behind me."

4 I borrow here from Emmanuel Levinas's conceptualization of a horizon of being, which represents a kind of limitless field of appropriation into the self. Levinas (2000, 9) explains how this way of viewing the world lends itself to abuses of beings who exist on the margins of that field: "In killing, I can certainly *attain* a goal, I can kill the way I hunt, or cut down trees, or slaughter animals – but then I have grasped the other in the opening of being in general, as an element of the world in which I stand. I have seen him on the horizon. I have not looked straight at him." The concept also appears in Levinas's *Totality and Infinity: An Essay on Exteriority* (1969, 43–4).

5 The indigenous woman who betrays her people by aiding colonists is a trope that refers back to Cortés and his interpreter and mistress, "La Malinche," who was first "gifted" to him. As Robert McKee Irwin (2010, 51) explains, "the rape of Mexico by Spain ... is typically represented allegorically through the figures of Hernán Cortés and La Malinche (aka Malintzin or Doña Marina) ... While on the one hand this representation of interracial relations implies the violence of armed conquest, with La Malinche being part of the spoils of war, on the other hand, she has traditionally been seen as a traitor to Mexico for collaborating with the enemy invaders." The narrator's suspicions and Teresa's interaction with the priest primarily through letters in *Winter in the Blood* refer back to language and the body as the vectors by which the native woman "betrays" her people.

6 Snakes have been a particularly common image, owing to Eve and the serpent Lucifer in the Garden of Eden. Lucifer's sins are pride against the rule of God, but also deviance in all its forms. When used against women, this deviance becomes particularly sexualized. This appears in American literatures throughout the nineteenth century. See Margaret Fuller's *Summer on the Lakes* (1844) and James Fenimore Cooper's novels, for example: in *The Prairie*, one man warns of the Sioux "devils, whose eyes are keener than the blackest snake's!" (1827, 39).

7 The Linnaean taxonomy of species here reflects the racist classification system of Johann Friedrich Blumenbach's Enlightenment-era racial science and phrenology, as well as the nineteenth-century Family Tree of Man. Anne McClintock (1995, 38) explains that "in the tree of time, racial hierarchy and historical progress became the fait accompli of nature ... [E]volutionary progress is represented by a series of distinct anatomical types, organized as a linear image of progress ... The entire chronological history of human development is captured and consumed at a glance, so

that anatomy becomes an allegory of progress and history is reproduced as a technology of the visible." Thus, phenotype rules all, and Evelina's visual metaphor becomes a particularly poignant point of resistance.

8 When Evelina writes that her inscription "held for [her] the sacred resonance of those Old Testament words written in fire by an invisible hand. Mene, mene, teckel, upharsin" (Erdrich 2008, 10). The sacred words she recalls from the Book of Daniel are both the names of currency and an omen of disaster for a nation – this confluence of commodity and a portent of disappearance evokes the worst of European colonialism in North America.

REFERENCES

Anderson, Eric Gary. 1999. *American Indian Literatures and the Southwest: Contexts and Dispositions*. Austin: University of Texas Press.

Archibald-Barber, Jesse Rae. 2009. "Trick of the Aesthetic Apocalypse: Ethics of Loss and Restoration in Thomas King's *Truth and Bright Water*." *Canadian Journal of Native Studies* 29 (1–2): 237–55.

Barthes, Roland. 1972. *Mythologies*, translated by Jonathan Cape. New York: Farrar, Straus and Giroux.

Blackburn, Carole. 2000. *Harvest of Souls: The Jesuit Missions and Colonialism in North America, 1632–1650*. Montreal: McGill-Queen's University Press.

Bruyneel, Kevin. 2007. *The Third Space of Sovereignty: The Postcolonial Politics of US-Indigenous Relations*. Minneapolis: University of Minnesota Press.

Byrd, Jodi A. 2011. *Transit of Empire: Indigenous Critiques of Colonialism*. Minneapolis: University of Minnesota Press.

Champagne, Duane. 2005. "Rethinking Native Relations with Contemporary Nation-States." In *Indigenous Peoples and the Modern State*, edited by Duane Champagne, Karen Jo Torjesen, and Susan Steiner, 3–23. Walnut Creek: AltaMira Press.

Christie, Stuart. 2009. *Plural Sovereignties and Contemporary Indigenous Literature*. New York: Palgrave Macmillan.

Conley, Robert J. 2005. *Cherokee Medicine Man: The Life and Work of a Modern-Day Healer*. Norman: University of Oklahoma Press.

Cooper, James Fenimore. 1898. *The Prairie*. New York: Houghton Mifflin.

Driskill, Qwo-Li. 2010. "Doubleweaving Two-Spirit Critiques: Building Alliances between Native and Queer Studies." *GLQ: A Journal of Lesbian and Gay Studies* 16 (1–2): 69–92.

Erdrich, Louise. 2001. *The Last Report on the Miracles at Little No Horse.* New York: HarperCollins.
– 2003. *Original Fire: Selected and New Poems.* New York: HarperCollins.
– 2008. *The Plague of Doves.* New York: HarperCollins.
Friedman, Susan Stanford. 1994. "Identity Politics, Syncretism, Catholicism, and Anishinabe Religion in Louise Erdrich's 'Tracks.'" *Religion and Literature* 26 (1): 107–33.
Fuller, Margaret. 1843. *Summer on the Lakes.* Boston: Freeman and Bolles.
Irwin, Robert McKee. 2010. "Lola Casanova: Tropes of *Mestizaje* and Frontiers of Race." In *Border Culture*, edited by Ilan Stavans, 50–92. Oxford: Greenwood.
Justice, Daniel Heath. 2006. *Our Fire Survives the Storm: A Cherokee Literary History.* Minneapolis: University of Minnesota Press.
King, Thomas. 1999. *Truth and Bright Water: A Novel.* Ontario: HarperCollins.
– 2003. *The Truth about Stories: A Native Narrative.* Toronto: House of Anansi Press.
LaDow, Beth. 2001. *The Medicine Line: Life and Death on a North American Borderland.* New York: Routledge.
– 2004. "Sanctuary: Native Border Crossings and the North American West." In *One West, Two Myths: A Comparative Reader*, edited by Carol Higham and Robert Thacker, 65–84. Calgary: University of Calgary Press.
Levinas, Emmanual. 1969. *Totality and Infinity: An Essay on Exteriority.* Pittsburgh: Duquesne University Press.
– 2000. *Entre Nous: Thinking-of-the-Other.* New York: Columbia University Press.
Lyons, Scott Richard. 2010. *X-Marks: Native Signatures of Assent.* Minneapolis: University of Minnesota Press.
Marx, Leo. 2000. *The Machine in the Garden: Technology and the Pastoral Ideal in America.* New York: Oxford University Press.
McClintock, Anne. 1995. *Imperial Leather: Race, Gender and Sexuality in the Colonial Contest.* New York: Routledge.
McCormack, Patricia A. 2005. "Competing Narratives: Barriers between Indigenous Peoples and the Canadian State." In *Indigenous Peoples and the Modern State*, edited by Duane Champagne, Karen Jo Torjesen, and Susan Steiner, 109–20. Walnut Creek: AltaMira Press.
Mooney, James. 1932. *The Swimmer Manuscript: Cherokee Sacred Formulas and Medicinal Prescriptions*, edited by Frans M. Olbrecht. Washington: United States Government Printing Office.

Newman, David, and Anssi Paasi. 1998. "Fences and Neighbours in the Postmodern World: Boundary Narratives in Political Geography." *Progress in Human Geography* 22 (2): 186–207.

Owens, Louis. 1998. *Mixedblood Messages: Literature, Film, Family, Place*. Norman: University of Oklahoma Press.

Pulitano, Elvira. 2003. *Toward a Native American Critical Theory*. Lincoln: University of Nebraska Press.

Ridington, Robin. 2006. "Happy Trails to You: Contexted Discourses and Indian Removals in Thomas King's *Truth and Bright Water*." In *When You Sing It Now, Just like New: First Nations Poetics, Voices, and Representations*, by Robin Ridington and Jillian Ridington, 288–311. Lincoln: University of Nebraska Press.

Ruppert, James. 1995. *Mediation in Contemporary Native American Fiction*. Norman: University of Oklahoma Press.

Sibley, David. 1995. *Geographies of Exclusion: Society and Difference in the West*. London: Routledge.

Teuton, Sean. 2001. "Placing the Ancestors: Postmodernism, 'Realism,' and American Indian Identity in James Welch's *Winter in the Blood*." *American Indian Quarterly* 25 (4): 626–50.

Turner, Frederick Jackson. 1920. *The Frontier in American History*. New York: Henry Holt and Company.

Vizenor, Gerald. 1997. *Hotline Healers: An Almost Browne Novel*. Hanover: Wesleyan University Press.

– 1998. *Fugitive Poses: Native American Indian Scenes of Absence and Presence*. Lincoln: University of Nebraska Press.

– 2005. "Crows Written on the Poplars: Autocritical Autobiographies." In *I Tell You Now: Autobiographical Essays by Native American Writers*, edited by Brian Swann and Arnold Krupat, 99–110. Lincoln: University of Nebraska Press.

Warrior, Robert. 1995. *Tribal Secrets: Recovering American Indian Intellectual Traditions*. Minneapolis: University of Minnesota Press.

Weaver, Jace. 2001. *Other Words: American Indian Literature, Law, and Culture*. Norman: University of Oklahoma Press.

Weaver, Jace, Craig S. Womack, and Robert Warrior. 2006. *American Indian Literary Nationalism*. Albuquerque: University of New Mexico Press.

Welch, James. 2008. *Winter in the Blood*. New York: Penguin.

West, Elliott. 2004. "Against the Grain: State-Making, Cultures, and Geography in the American West." In *One West, Two Myths: A Comparative Reader*, edited by Carol Higham and Robert Thacker, 1–22. Calgary: University of Calgary Press.

9

The Anishnaabeg of Bawating: Indigenous People Look at the Canada-US Border

PHIL BELLFY

INTRODUCTION

Bawating is the Ojibwe name for the area at the mouth of Lake Superior, and can be translated as "the gathering place of the people." That same area is a place now called Sault Ste Marie – French for "the Rapids of the St Mary's River." These "Twin Saults" are divided by the Canada-US border, which puts one "Sault" in Ontario and the other in Michigan. This very same border divides the Anishnaabeg of the region; there are two tribes on the southern side of the border, and two First Nations to the north.

The Indigenous people of the region refer to themselves as *Anishnaabeg*, which translates to "the people who intend to do well." They are also called the *Chippewa*, a term used exclusively in the United States, and the *Ojibwe* (with various spellings), a term used exclusively in Canada. These two different, politically tinged terms show how the visitor governments attempted to separate these sovereign people into competing and distinct camps and used a divide-and-conquer strategy as they drew the Canada-US border through the middle of their community. While this division has been historically successful to a large degree, the Anishnaabeg of the Bawating region are in the process of breaking down that border. In this essay, I will attempt to do the same thing: break down the border and make it less than the solid line that most people accept today.

FIFTEENTH-CENTURY PAPAL BULLS AND THE ANISHNAABEG

We grant you [Kings of Spain and Portugal] by these present documents, with our Apostolic Authority, full and free permission to invade, search out, capture, and subjugate the Saracens and pagans, and any other unbelievers and enemies of Christ wherever they may be, as well as their kingdoms, duchies, counties, principalities, and other property ... and to reduce their persons into perpetual slavery.

Nicholas V 1452

While it may seem odd to those who accept the inviolability of the current Canada-US border to see a reference to a fifteenth-century papal bull at the start of this discussion, fifteenth-century history – and how it unfolded in the upper Great Lakes – is critical to understanding contemporary Anishnaabeg views of that border. This is due to the simple fact that underlies the *Dum Diversas* papal bull, and that is the arrogance of the European "explorers" – Columbus included – who felt empowered by the Pope to do as they liked with the "pagans" they encountered and their "kingdoms, duchies, counties, principalities, and other property" (Nicholas V 1452).

In 1493, after the *Dum Diversas* and necessitated by the "discovery" of the Americas, Pope Alexander VI issued the papal bull *Inter Caetera*. This bull established a line "from the Arctic pole ... to the Antarctic pole" that granted Spain "all islands and mainlands found and to be found, discovered and to be discovered" as long as any of those lands had not yet been "acquired by any [other] Christian prince." Together, these two bulls – as well as a third, the *Romanus Pontifex* – "regulated" land seizures and slavery in the so-called Age of Discovery, and essentially gave monopoly rights to one European power over any others that might also be inclined to seize lands.[1]

On 14 June 1671, the French held the Pageant of Saint Lusson in Bawating on the north side of the river. Through this pageant, with many "dusky savages" in attendance, the French laid claim to all of North America. Lusson declared:

> In the name of the most high and redoubtable sovereign, Louis the Fourteenth, Christian King of France and Navarre, I now take possession of all of these lakes, straits, rivers, islands, and regions lying adjacent thereto ... and I declare all of the people

inhabiting this wide country that they now become my vassals [and that] other princes and potentates of whatever rank ... are denied forever seizing upon or settling within these circumjacent seas. (Reuben 1883)

The reference to the Christian King of France, and his royal warning that banned "other princes and potentates" from claiming any of these lands, echoes the fifteenth-century papal bulls that legitimized earlier land claims. Lusson's bull-like declaration, embellished by the rhetorical flourishes of the time, was most likely lost on the Native people the French had assembled and meant to impress with such pageantry. But it is unmistakable that Lusson was directing his pronouncement less to the Indigenous people who had been called to the pageant than to the princes and potentates of England, Spain, and any other European power that might have sought to assert its own sovereignty over the vast continent.

It is not a coincidence that the language of this 1671 pageant is similar to the papal bulls, even though their pronouncements were separated by two centuries. All documents were intended to establish a formal mechanism whereby one European state could make a claim to what it called pagan lands and, backed by the force of arms, hope to persuade other powers to honour those claims. These declarations – including those made by Columbus – provided the European powers with legal self-justification.

COLONIAL ERA PROCLAMATIONS

In 1763, the British government issued a royal proclamation that claimed it had a monopoly over the disposition of "French" lands in North America. That is, the proclamation laid formal claim to much of the same land that France had claimed in the Sault in 1671 and lost to Great Britain after The Conquest of 1760. The Royal Proclamation declared that the lands west of the Appalachian Mountains were "Indian territory" and therefore closed to colonial expansion. Its language is clear on this point:

> And whereas it is just and reasonable, and essential to our Interest, and the Security of our Colonies, that the several Nations or Tribes of Indians with whom We are connected, and who live under our Protection, should not be molested

or disturbed in the Possession of such Parts of Our Dominions and Territories as, not having been ceded to or purchased by Us, are reserved to them, or any of them, as their Hunting Grounds.

There are two important points in this proclamation: first, it recognizes that the Indigenous people of the region have a sovereign right to their "Hunting Grounds." Second, it recognizes that any such lands "in the Possession" of "Indians" can only be ceded to or sold to the Crown, and only if the "Indians" are inclined to do so – that is, they cannot be "molested or disturbed." To strip aside the formal eighteenth-century language of diplomacy, the proclamation echoes the formal seventeenth-century language of the Pageant of Saint Lusson, but directs it more to the new American government than to the European powers that might challenge British sovereignty. As a result, in 1763, France and Britain agreed that formerly "French" land was "British" – though they also admitted that much of North America was really Indian territory.

This closing of Indian territory to colonial expansion was an important underlying factor that led to the American Revolution; yet, in 1783, the newly established United States of America issued a proclamation that declared its government was now the sovereign power in Indian territory, and that it alone had the power to obtain cessions of land from the "Indians" (or the right to purchase it from them). In 1787, the US federal government passed the Northwest Ordinance. The relevant passage reads: "The utmost good faith shall always be observed towards the Indians; their land and property shall never be taken without their consent; and, in their property, rights, and liberty, they shall never be invaded or disturbed." Again, it is important to note that the US government recognized Native peoples' sovereign rights to their "land and property." Also, as did both the royal and the federal proclamations, the US government maintained itself as the entity with the sole authority to treat with Indigenous people for their lands.

AFTER THE AMERICAN REVOLUTION

All of these bulls, pageants, proclamations, and ordinances shared a dominant theme: in each, all "civilized" nations recognized the right of one European power to steal, trade, or buy the lands of Indigenous people – even if that one power were to change. Of course, after the

American Revolutionary War, all of this had to be sorted out, once again, on the international scene. The 1783 Treaty of Paris did just that for the new US and British governments after the revolution. The indigenous people of the region recognize that the boundary the treaty established between the United States and Great Britain was not much more than a formal agreement as to which side was authorized to "treat" with which Indians over the "dispossession" of lands that constituted Indian territory.

In its discussion of the boundary between the two countries in the Bawating area, article 2 of the Treaty of Paris states: the boundary will be drawn "along the middle of said water communication into Lake Huron, thence through the middle of said lake to the water communication between that lake and Lake Superior." The St Mary's River – the water communication – is full of islands, which also leads to a number of problems. The rather ambiguous language that defined the boundary between the two countries created a situation that also threatened the "firm and perpetual peace between his Brittanic Majesty and the said states, and between the subjects of the one and the citizens of the other" (article 7). To resolve many of the issues left unsettled by the Treaty of Paris, the US government sent Supreme Court Justice John Jay to London to negotiate the Treaty of Amity, Navigation, and Commerce between His Britannic Majesty; and the United States of America, which the two governments ratified in 1794. Given that it has a rather cumbersome title, the treaty is more commonly referred to as the Jay Treaty.

While the Treaty of Paris does not mention Native people, article 3 of the Jay Treaty does contain a significant reference:

> It is agreed that it shall at all times be free to His Majesty's subjects, and to the citizens of the United States, and also to the Indians dwelling on either side of the said boundary line, freely to pass and repass by land or inland navigation, into the respective territories and countries of the two parties, on the continent of America, (the country within the limits of the Hudson's Bay Company only excepted.) ... No duty of entry shall ever be levied by either party on peltries brought by land or inland navigation into the said territories respectively, nor shall the Indians passing or repassing with their own proper goods and effects of whatever nature, pay for the same any impost or duty whatever.

This article makes explicit the political independence and sovereignty of the Indigenous people of North American Indian territory. That is, it defines three separate groups of people: British subjects, American citizens, and "Indians dwelling on either side of the boundary." Of course, the article does not address the problem of determining just where that boundary lies. In fact, article 4 of the treaty simply kicks the issue down the road and stipulates "the two parties will thereupon proceed, by amicable negotiation, to regulate the boundary line in that quarter [Lake of the Woods], as well as all other points to be adjusted between the said parties, according to justice and mutual convenience, and in conformity to the intent [of the Treaty of Paris]."

The War of 1812 broke the "firm and perpetual peace" – the "Amity" the Jay Treaty refers to – long before the surveyors and commissioners could figure out just how to adjust all those other points of the boundary that the parties had yet to agree on. The 1814 Treaty of Ghent formally ended the War of 1812, and includes language relevant to the indigenous perspective in article 9:

> The United States of America engage to put an end immediately after the Ratification of the present Treaty to hostilities with all the Tribes or Nations of Indians with whom they may be at war at the time of such Ratification, and forthwith to restore to such Tribes or Nations respectively all the possessions, rights, and privileges which they may have enjoyed or been entitled to in one thousand eight hundred and eleven previous to such hostilities.

Later in the same article, Great Britain agrees to the same restoration of rights, including all the "possessions, rights and privileges" laid out in the Jay Treaty. The Treaty of Ghent also recognizes the indeterminate nature of the border between the two countries, as article 6 refers to the problem of the water communication discussed earlier:

> And whereas doubts have arisen what was the middle of the said River, Lakes, and water communications, and whether certain Islands lying in the same were within the Dominions of His Britannic Majesty or of the United States: In order therefore finally to decide these doubts, they shall be referred to two Commissioners to be appointed, sworn, and authorized to act [to resolve these issues].

Despite the good intentions of the parties laid out in the 1814 Treaty of Ghent, the survey of the boundary through the length of the St Mary's River did not begin until 1828, and, at that time, was not completed due to a disagreement as to just which country would be awarded Sugar Island, the northernmost island in the St Mary's. In 1942, when the line was finally determined, the commissioners refer to that issue as follows: "From the place where the joint Commissioners *terminated their labors* under the sixth article of the Treaty of Ghent, to wit: at a point in the Neebish Channel, near Muddy Lake" (emphasis added). The commissioners "terminated their labors" in 1828.

It may be time for a brief timeline of the Canada-US boundary situation as it relates to the upper St Mary's River of Bawating:

1787   The Northwest Ordinance recognizes the sovereign and territorial rights of the native peoples of Indian territory (which the ordinance refers to as the "Northwest").

1794   The Jay Treaty recognizes the distinct and sovereign independence of the area's native people.

1814   The Treaty of Ghent reaffirms those sovereign rights, and sets up a process to determine which country has the right to treat with native people for their land.

1828   Treaty of Ghent commissioners cannot decide where the border should be drawn through the upper St Mary's River and are divided over who should get Sugar Island (St Georges Island). The "termination line" drawn by the surveyors also leaves Neebish Island in that same "indeterminate" state.

1842   The boundary between the United States and Great Britain in the upper St Mary's is finally settled, and Neebish Island and Sugar Island fall on the US side of the border.

In 1842, the outstanding border issues between Great Britain and the United States were finally settled, but the resolution of the border issue did not give these islands to the United States. Instead, it only gave the United States the right to negotiate with the area's Native people for a cession of Neebish Island and Sugar Island (along with its islets). At this point, the Indian territory proclaimed by both the United States and Great Britain in the Upper Great Lakes was still under the absolute sovereignty of the area's Indigenous people. Not one acre of it had been ceded, sold, or lost in a "just war."

LINES DRAWN UPON THE WATER[2]

Despite all surveyors' maps, agreements, and the formal language of various treaties and proclamations, the border situation in the upper St Mary's was far from resolved. To set aside the issue of the disposition of Neebish and Sugar Islands for a moment, the surveyors simply drew lines down the middle of the "water communication" that then became the official boundary. But lines drawn upon the water can never be firmly established, as water erodes some shores, wind and sand build up others, and rising and falling water levels can dramatically change where the middle of the channel falls. One modern example illustrates the quite literally fluid nature of the boundary in the area.

On the surveyors' nineteenth-century maps, the lines flow, much like the waters they are drawn on, down the middle of each river or channel. The various treaties and agreements demand such a division. However, more modern surveyors have felt the need to draw each segment of the border from fixed point to fixed point as a straight line with very precise angles. The result is a border that becomes a series of firmly determined straight lines, and if these more "modern" surveyors try to keep the number of segments to the smallest possible, sometimes these straight line segments are not all that close to the centre of the river.

This is what has happened to the border at the far north end of Sugar Island. The St Mary's water level has dropped to such a degree that portions of "Canadian" land that used to be submerged now appears to have emerged on the American side of the boundary (see figure 9.1). To complicate things even more, the "US land" shown on the map is now, without question, territory of the "Canadian" Garden River First Nation. The indeterminate status of the Canada-US boundary leads to a whole series of questions about the "inviolable" nature of the "world's longest undefended border" or "longest secure border" after the affairs of 9/11.

THE INDETERMINATE NATURE OF THE BORDER AND THE 9/11 ATTACKS

The physical Canada-US border is lined with uniformed guards with badges and guns. There are immigration and customs facilities, posts or pillars, fences, barriers, gates, X-ray scanners, Geiger counters, drug- and bomb-sniffing dogs, and fleets of SUVs. All of this security paraphernalia lends an air of permanence and inevitability to "the

Figure 9.1 Satellite image that shows the Canada-US border as drawn between Garden River First Nation and Sugar Island (Google 2012). The land south of the line is clearly Garden River territory.

border," as if it was something very real. But, the reality of the situation is not what it appears to be – especially as the border affects native people. All of this security is relatively new, and most of it was instituted in the wake of the terrorist attacks of 9/11 (setting aside the fact that not one of the nineteen bombers crossed the Canada-US border in furtherance of those attacks [see Struck 2005]). Prior to 9/11, Native people on either side of the border enjoyed unfettered crossing (other than having to pay a bridge toll, if driving) if they had a tribal card to serve as valid identification. 9/11 transformed a stress-free crossing into one fraught with tension – even the Canadian border guards are now armed and demand passports, and a hint of apprehension on the part of a traveller can lead to detention, a search, and a half-hour delay on top of the sometimes-two-hour delay just to get to the guard station that has become normal at Sault Ste Marie and other busy entry points.

Increased border security affects all travellers, not only Indigenous people. But although most people, including most border guards, are hesitant to recognize it, Indigenous people do enjoy "special rights" as they cross and re-cross the border, rights that the language of the Jay Treaty recognizes. Historically, Native people have often invoked article 3 of the Jay Treaty to bring goods into Canada (or to a lesser degree, into the United States) without paying "any impost or duty whatever." But, with a frustrating consistency, the Canadian courts have denied Native people this treaty right and claim it has never been enacted by legislation by the 1867 post-confederation Canadian government, setting aside the implications of both Upper Canada and Lower Canada enacting exactly such duty-free importation legislation before confederation. The most recent duty-free Canada Supreme Court ruling – in a case brought by Chief Mike Mitchell of Akwesasne, Native land between New York, Ontario, and Quebec, and between the United States and Canada – was also denied, despite the fact that section 35 of the Constitution Act of 1982, "recognizes and affirms the existing aboriginal and treaty rights." One must assume (as Native people did and do) that this constitutional provision applies to those rights enshrined in the Jay Treaty.

When First Nations people look at that "not enacted by legislation" argument against the exercise of their Jay Treaty rights, they can now point to recent research by Dr David McNab in their defence. McNab, a land claims researcher, has concluded that the Canada-US border has never been enacted by legislation, either (see McNab 2004, 33–4). From the Native perspective, it is more important to note that the border has not been enacted through legislation than to worry about the duty-free implications. In other words, all of the post-9/11 border security measures have been implemented in defence of an imaginary line drawn on a map, albeit a line that both the United States and Canada have accepted as fact, simply because both countries assume the border has been drawn in accordance with precise treaty language, even though that language has been modified several times.

To move far from Bawating, the most recent example of the "fluid" nature of the Canada-US border concerns boundaries regarding sovereignty in the Arctic. As the polar ice melts, and as vast areas become open to mineral and oil development, the issue of "who owns what" has become serious as the United States, Canada, Russia, Greenland (Denmark), and other states struggle to assert jurisdiction over what they claim is theirs (the Russians literally planted a flag on the seabed floor in 2007 and touched off an international firestorm). More recently, in August 2010, the United States and Canada faced off over the interpretation of the 1825 Treaty of St Petersburg between Russia and Great Britain concerning which country owns what in the Arctic Sea just off the coast between Alaska and the Northwest Territories.

THE 1836 TREATY OF WASHINGTON AND THE 1850 HURON-ROBINSON TREATY

The implications of this Alaska-NWT border "line drawn upon the water" dispute is quite germane to the issues currently under dispute in Bawating. In 1836, the US government and the area's indigenous people signed the "land-cession" Treaty of Washington. Through this treaty, the US government agreed that "Sugar Island, with its islets ... shall also be reserved for the use of the Chippewas living north of the straits of Michilimackinac" (article 3). The problem with this statement is that the border in the area had not yet been determined, and so Sugar Island was not under US jurisdiction. That is, the United States did not – by treaty or any other legal instrument – have an agreement with Great Britain that gave either country the right to negotiate with the region's Native people over the disposition of their lands.

The 1836 Treaty of Washington refers to the St Mary's River boundary as it describes the limits of the land it claims is being ceded to the United States as follows: "Thence northeast to the boundary line in Lake Huron between the United States and the British province of Upper Canada, thence northwestwardly, *following the said line, as established by the commissioners* acting under the treaty of Ghent, through the straits, and river St Mary's, to a point in Lake Superior north of the mouth of Gitchy Seebing" (article 1, emphasis added). When looking at this border, it is important to note that the line established by the commissioners terminated at a point south of Sugar and Neebish Islands, a line that did not continue until that point "in the middle of the St Mary's river, about one mile above St George's or Sugar Island." As a result, the northern terminus of the disputed area falls just below the rapids (see figure 9.2).

Further muddying these waters, in 1837, in advance of Michigan becoming a state, Henry Rowe Schoolcraft, the area's Indian agent, drew up a map and census of all of Michigan's reservations and the "number of [Indian] souls" that resided on each. While Sugar Island appears on the map as a reservation, the accompanying census data has no reference to it. Given the indeterminate international status of both Sugar and Neebish Islands due to the commissioners' failure to firmly establish the border in this area, and Schoolcraft's implicit recognition of this indeterminate status through his failure to include the "souls" of Sugar Island in his Michigan census, Native people of the area believe that Sugar and Neebish Islands (and their islets)

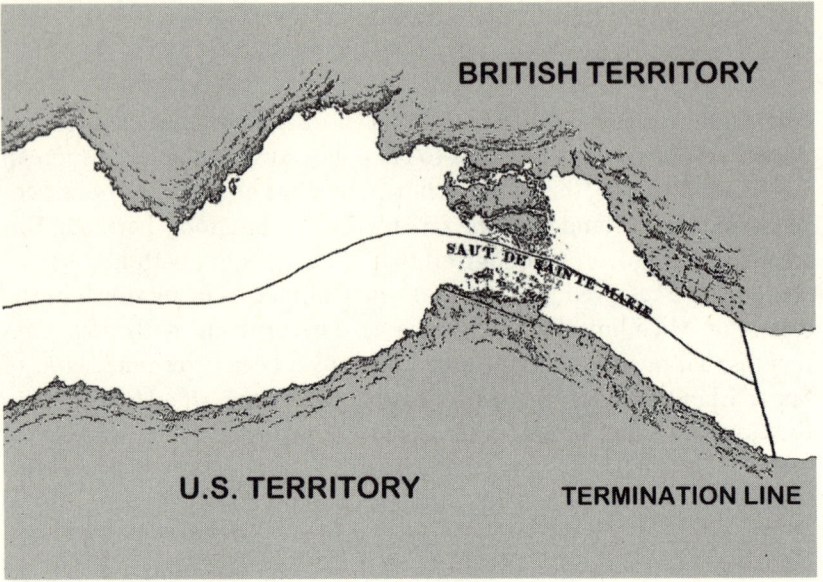

Figure 9.2 Treaty of Ghent boundary, upper termination line, 1828 (Moore 1898)

have yet to be "ceded" to anyone. Consequently, these islands cannot be said to be a part of either the United States or Canada.

In 1836, when the United States gained a cession of considerable territory in what is now northern Michigan, the entire border area between the two termination lines – including Neebish and Sugar Islands and its islets – drawn by the Treaty of Ghent commissioners was not under US jurisdiction. That is, the United States and Great Britain could not agree who had the right to treat with the native people of the area for a cession of that portion of Indian territory, a formal status accorded to those "Hunting Grounds" under several treaties and agreements discussed above.

Land north of the border in this area was subject to the land cession provisions of the 1850 Huron-Robinson Treaty. Given that the status of the upper St Mary's border was agreed on in 1842 under terms of the Webster-Ashburton Treaty, which gave the United States the right to treat with the Native people of the area, Canada did not include this territory in its 1850 treaty; that is, the Crown did not obtain a cession of Neebish and Sugar Islands, either.

In 1836, the border commissioners did not determine the status of Neebish and Sugar Islands, and so the United States could not have

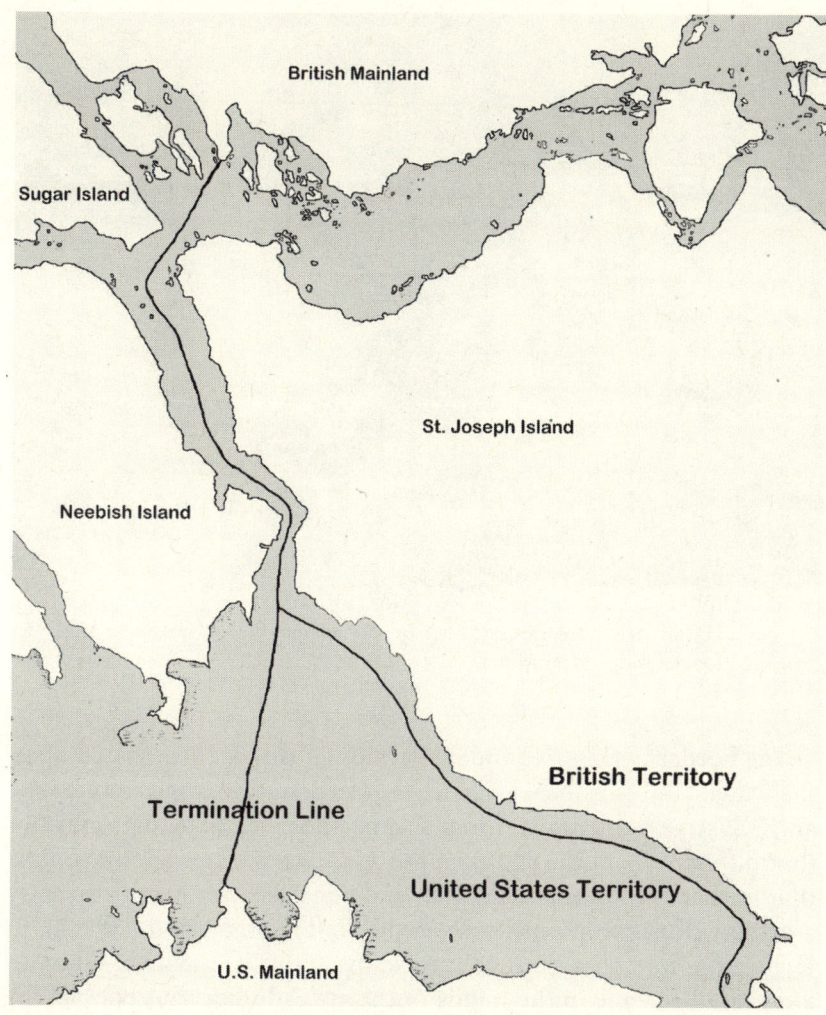

Figure 9.3 Treaty of Ghent boundary, lower termination line, 1828 (Moore 1898)

obtained its cession through the Treaty of Washington. In 1850, the British also did not negotiate a cession of this territory. The question I will now ask is: when, and through which international treaty, did the Indigenous people of the region cede this territory? The indeterminate status of the area at the time of these two treaties, in 1836 and 1850, is evidence that it could not have been ceded to either country according to any treaties or agreements recognized by either or both.

Figure 9.4 Detail of Schoolcraft's 1837 map of Michigan that shows Sugar Island as "reserved for the Indians" (author)

The border in this area today features a distinct "line drawn upon the water." But this line is subject to interpretation, especially by the area's Native population. From this perspective, the situation is not that different from the status of the Alaska-NWT border. But unlike that border, the status of Neebish and Sugar Islands is not currently a subject of negotiations between the United States and Canada. In fact, in the minds of both countries, the status of the area is not an issue at all – while in the minds of the area's Indigenous people, the situation is far from resolved.

TREATY RIGHTS AND THE BORDER AFTER 9/11

This historic, indeterminate status of the border as it wended its way through the upper St Mary's River over a sixty-year period is more than a passing curiosity. Interest in Sugar Island by the area's Indigenous people has intensified since the US government beefed up its border security through the Western Hemisphere Travel Initiative, and other initiatives, in the wake of 9/11. The argument for unfettered crossing, especially for First Nations citizens north of the

A Map of the Acting Superintendency of Michigan Population			
	souls		
Sault Ste. Marie Bands	180	White river	142
Tacquamenon river	77	Maskigo	94
Drummond Island	64	Grand Traverse B.	417
Grand Island	66	Little Traverse B.	497
River aux Traines	2	L'Abre Croche	314
Chocolate River	73	Village of the Cross	225
Esconawba river	111	Rain's Band	164
Shawan Egeezhig's band	127	Fort Village G.R.	118
Little Bay de Nocquet	109	" " "	38
Beaver Islands	117	Little Prairie "	53
Bear Skin's Band	108	Grand Rapids "	160
Ance & Missatigo's	157	Prairie Village "	47
Chenos	75	Thornapple R. "	106
Michilimackinac & Bois Blanc	72	Forks "	76
Cheboigan	112	Flat River "	135
Thunder Bay	109	Maple River "	156
Carpe river	138		Souls 4561
Plate river	9	Estimated number of Chippewas in	
Manistee river	45	Michigan west of the cession of 1836	1200
Pierre Marquette	68	Menomonees between Esconawba &	
	1819	Menomonee rivers. Estimate.	60
		Pottomattomies & Chippewas & Ottawas	
		South of Grand River estimate.	500
		Saganaws of Michigan "	1000
		Swan Creek & Black river Chippewas	300
		Total population within the limits of Michigan	7621
		Ottawas of Maumee, in Ohio estimate	200
			7821

Figure 9.5 Transcription of a portion of Schoolcraft's 1837 map of Michigan that shows a census of Michigan Indians. Sugar Island is not listed (author).

border in the Sault area, is now focused on the Jay Treaty language that guarantees "the Indians dwelling on either side of the said boundary line, [can freely] pass and repass by land or inland navigation, into the respective territories and countries of the two parties, on the continent of America" (article 3).

The argument has two major points: first, the US Western Hemisphere Travel Initiative *legislation*, enacted in 2004, cannot replace *treaty* language. That is, the US constitution declares that treaties are the "Supreme Law of the Land" (6.1.2); therefore,

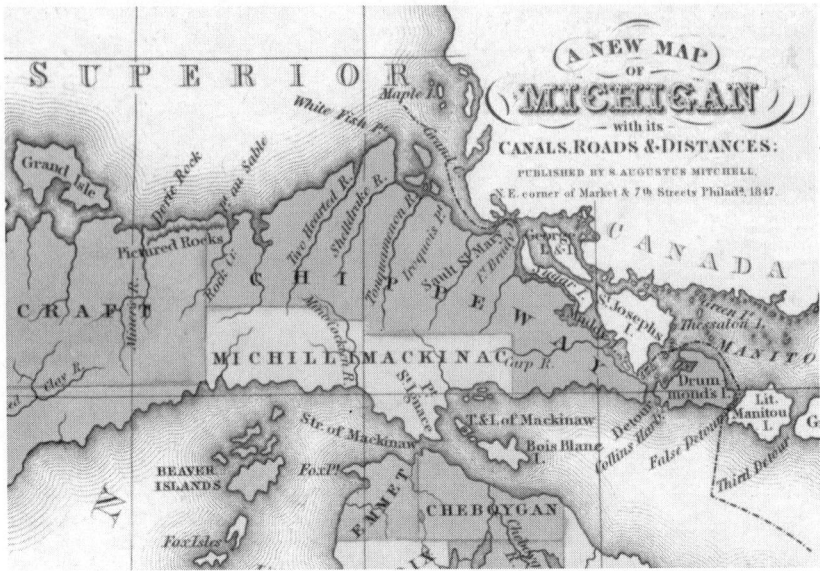

Figure 9.6 Portion of the 1847 Mitchell map of Michigan that shows Sugar Island as part of Canada

Congress cannot pass a law that restricts any rights that treaties have heretofore guaranteed. Secondly, and based on that treaty right, First Nations people do not agree that they cross the border when they drive over the bridge from Sault, Ontario, to Sault, Michigan. They have right to unfettered travel throughout the area, since Batchewana, Garden River, Sault Tribe, and Bay Mills territory is neither US territory nor Canadian territory – it is Anishnaabeg territory and is recognized as such by treaty, not legislation.

The Batchewana, Garden River, Sault Tribe, and Bay Mills tribes compose the Anishnaabeg Joint Commission (AJC),[3] formed in the aftermath of extensive raw sewage contamination along the northern shore of Sugar Island in the summer of 2006. It was determined that the source of the contamination came from the Sault, Ontario, sewage treatment plant, which was undergoing an upgrade at the time. The sewage apparently came from Canada, but washed up in the United States, and so neither government wanted to get involved and each claimed it lacked jurisdiction. The four tribes of the AJC stepped into this vacuum and signed a treaty that committed them to assert their collective jurisdiction over the entire St Mary's watershed. As a consequence, tribal members became responsible for protecting and preserving the upper St Mary's River through, for example, collecting water samples, with no concern over border jurisdiction issues.

SUGAR ISLAND AND INTERNATIONAL INDIGENOUS JURISDICTION

The Sault Ste Marie Tribe of Chippewa Indians and the Bay Mills Indian Community both have modern reservation lands on Sugar Island, though they may now argue that the island is an international reservation because it was not ceded in the 1836 Treaty of Washington (with the United States) or in the 1850 Huron-Robinson Treaty (with Great Britain). Furthermore, the 1855 allotment Treaty of Detroit, signed by the United States with the Ottawa and Chippewa – that is, the Treaty of Washington signatory tribes – contains the following language: "The benefits of [article 1] will be extended only to those Indians who are at this time actual residents of the State of Michigan, and entitled to participate in the annuities provided by the treaty of March 28, 1836; but this provision shall not be construed to exclude any Indian now belonging to the *Garden River band* of Sault Ste. Marie" (emphasis added). The "Garden River band of Sault Ste. Marie" referred to people in Sault Ste Marie, Ontario, a fact of which the United States was aware. The 1855 Treaty of Detroit also had Native signatories from both sides of the river, that is, from both sides of the border, a fact that was also clearly known to the US treaty negotiators (Bellfy 2011).

In another "cross-border" recognition of the fluidity – or absence – of the pre-9/11 Canada-US border, citizenship in the Sault Ste Marie Tribe of Chippewa Indians is open to descendants of anyone listed on the "Garden River Annuities (1846–1925) and Garden River Church Records," among other rolls (Sault Tribe of Chippewa Indians 2012a). This provision is delineated in the Sault Tribe Constitution in article III, section 1, which was ratified in 1975. Obviously, citizenship in the Sault Tribe is not related to any issues of "borderland residency." In fact, the tribe has origins on Sugar Island, and is but one more example of the common heritage, culture, and history the entire region's Indigenous people share without regard to the border (Sault Tribe of Chippewa Indians 2012b).

It should be noted that the "visitor" governments have always formally recognized the sovereign status of the area's Indigenous people without reference to residency – that is, whether they live in the United States or in Canada – a recognition that spans the centuries since first contact (which, for the Bawating area, was French contact in the mid-1600s).

Other recent developments have cemented that "cross-border" Indigenous relationship. As I discussed above, a 2006 area-wide treaty affirmed Indigenous jurisdiction over the entire St Mary's River watershed, which led to the establishment of the Anishnaabeg Joint Commission. In 2008, its same four signatories – the Bay Mills Indian Community, the Sault Ste Marie Tribe of Chippewa Indians, the Batchewana First Nation, and the Garden River First Nation – signed the 2008 Summit Treaty.[4] This treaty reaffirmed their common history, territorial integrity, and sovereignty over the area's entire "traditional territory." The treaty also demanded that both the US and Canadian governments recognize tribal members' rights to travel unimpeded throughout those traditional homelands without regard to the Canada-US border.

After signing two copies of the treaty on Sault Tribe territory, the four tribal representatives (and others) travelled by canoe from Sugar Island to Garden River First Nation territory, directly across the St Mary's River, and signed two more copies. It should be noted that residents of Sugar Island enjoyed unrestricted travel rights to Garden River – and vice versa – before border security intensified after the events of 9/11.

This cross-border canoe travel took place once again in August 2010; this time, the human remains of six Batchewana Natives were repatriated from the Smithsonian Institution in Washington, DC. In 1867, the US war department surgeon general issued a circular that requested "medical officers and others ... forward to the Army Medical Museum ... typical crania of Indian tribes, specimens of their arms, dress, implements ... etc" (Barnes 1867). Army surgeon Joseph H.T. King, stationed at the US Fort Brady, was one of those who answered this call and sent any number of human remains to the Army Medical Museum. Of those, the Smithsonian Institution, which took control of the remains in 1898, recently identified six as "culturally affiliated" with the Batchewana First Nation (Hollinger et al. 2009).

King, in a letter dated 7 October 1875 sent from Fort Brady, states that "through the kindness of my friend Capt. Wilson, who resides in Canada opposite [Fort] Brady [I received] the crania of a full-blooded Cree Indian." Another of King's letters, dated 10 August 1875, outlines one of his excursions to the north shore of Lake Huron to the appropriately named Dead Island, a Native burial place King refers to as "my hunting ground." This island was situated some 150 miles east of Fort Brady, deep in Canadian territory. King sent the human remains repatriated to Batchewana in 2010 to the Army Medical Museum in June of 1875. These six remains were not any of those referred to by King in either of the two letters referenced above; it appears that King was very busy in the summer of 1875.

Such incidents feature a number of notable cross-border details. First, the US army surgeon general's request for human remains was not interpreted as restricted to US territory. What's more, as King's correspondence shows, Canadian military personnel, ostensibly stationed on the border to prevent incursions by US military into Canada, not only ignored military grave-robbers crossing into Canadian territory, but also acted alongside them. The only adequate explanation for Canada's lack of concern may lie in the language of the Jay Treaty; that is, the human remains dug up in Canada and shipped to Washington were not considered "Canadian Indians," but simply "Indians dwelling on either side" of the Canada-US boundary. That is, if they were not "Canadian Indians," no violation of Canadian sovereignty occurred when US army officers came to Canada to dig them up – even their burial sites can be considered Indian territory, not Canada proper.

The fact that a US army surgeon was "bone-harvesting" across the upper Great Lakes with the assistance of Canadian military officers is significant in and of itself, but more notable is that the return of these remains to Batchewana First Nation marked the first time the Smithsonian worked with "American" tribes to conclude a repatriation of human remains to a "Canadian" First Nation (*Sault Star*, 20 August 2010). This first required the Smithsonian to craft the legal arguments for such repatriation as the enabling legislation – the Native American Graves Protection and Repatriation Act (NAGPRA) – only authorizes the museum to repatriate human remains to US-government recognized tribes. To meet this requirement, the Smithsonian characterized the remains as "potentially affiliated with the Sault Ste. Marie Band of Chippewa Indians and the Bay Mills Indian Community" (Hollinger et al. 2009, 2). As Smithsonian officials explained, once they repatriated the remains to a US federally recognized tribe, they had fulfilled their obligations under the law; what happened to the remains after that was not an issue of concern (Hollinger, pers. comm.).

The entire Batchewana repatriation process was initiated by a Smithsonian representative during a 2002 museum visit by Sault Tribe and Bay Mills members, who were in Washington to oversee the repatriation of human remains to their communities. During that meeting, the Smithsonian representative mentioned that the institute had human remains that were from Ontario. In the following year, the Smithsonian began discussions with several First Nations, including Batchewana, in order to more fully identify the "Chippewa" group that could be most closely linked to the remains. Over the course of the next few years, it was determined that the remains were most closely affiliated with Batchewana, and the discussions that ensued included all four of the area tribes, that is, all four AJC tribes. Hollinger and others (2009, 3) characterize the underlying history as follows:

> Historically, the Anishnaabeg inhabiting the Sault Ste. Marie region moved freely on both the Canadian and US sides of the border for hunting, gathering, and trading, but they regularly congregated at the rapids of the St Mary's River to fish and interact. The fluidity of the border for these people continues to exist today as Anishnaabeg from Sault bands intermarry and perpetuate a common identity as "people of the rapids." The descendants of the Anishnaabeg of the Sault Ste. Marie region today are found among the Sault Ste. Marie Band of Chippewa Indians and the Bay Mills Indian Community as well as the First Nations at Batchawana and Garden River.

Of course, the US government's recognition of the "fluidity of the border" is a refreshingly frank acknowledgement of what the area's Indigenous people have known and understood for centuries; nevertheless, it is good to have a US government official make such an honest statement, even though it has taken most of those same centuries to do so.

INDIGENOUS NATIONS OF NORTH AMERICA AND THE CANADA-US BORDER

There is one last event to recount concerning regional indigenous people and this newly recognized fluidity of the border. For that, I will return to the Western Hemisphere Travel Initiative (WHTI), the post-9/11 legislation mentioned above. In recognition of the cross-border practices of Native people, the US federal government put provisions into the initiative to modify travel restrictions placed on non-Native travellers and specifically outlined tribal alternatives to standard government-issued passports. According to WHTI, all of these tribal passport alternatives would require that the issuer – the First Nation or the tribe – declare the "citizenship" of the bearer; that is, state that the bearer was either an "American citizen" or a "Canadian citizen." Such a designation was an affront to the sovereignty of tribal people.

In an attempt to resolve the identification issue, the AJC engaged in discussions with US Department of Homeland Security officials in Washington, DC. The First Nations representatives were adamant that they would never agree to call themselves "Canadian citizens." What is curious about all of this is that these discussions had to take place between homeland security officials and representatives of US federally recognized tribes, mirroring the repatriation discussions between the tribes, First Nations, and Smithsonian officials. (The Sault Tribe and Bay Mills were involved in these discussions on behalf of their fellow AJC tribes, Batchewana and Garden River.)

On 1 July 2010, the US Department of Homeland Security and tribal officials signed a memorandum of agreement (MoA) that set out the parameters for Batchewana and Garden River First Nations' "enhanced tribal identification card" in order for it to be accepted by US border officials. The issue of declaring one's citizenship was resolved by the MoA through the inclusion of the following language: "The following is a listing of treaties and legislation that show the relationship between Batchewana First Nation, Garden River First Nation, Bay Mills Indian Community, and Sault Tribe as alliances, and as Indigenous Nations of North America." Included in that list are the 1836 and 1855 United States treaties and the 1850 Huron-Robinson Canadian treaty mentioned in this paper. The basis for this statement is that those who signed these treaties did not sign as "American Indians" or as "Canadian Indians" but as members of the "Anishnaabeg Nation," a fact that this memorandum recognizes (Batchewana First Nation et al. 2010).

The United Nations (1945, 1.2.4) describes a sovereign nation as one that possesses "territorial integrity and political independence." Canada and the United States now recognize that the Indigenous people of Bawating have both of these attributes. They have demonstrated their political independence by signing numerous treaties and international agreements. In that July 2010 memorandum of agreement, the US government recognized – for the first time in history – that the Anishnaabeg are neither Canadians nor Americans, but are citizens of an "Indigenous Nation of North America" – Bawating, or the Gathering Place of the People. This document lists several international treaties and agreements that also recognize Bawating's territorial integrity and occupation of the region since time immemorial. Bawating is more than a place and more than a region on a map – it is a statement of identity and culture and sovereignty and jurisdiction.

On any map, Bawating's territorial integrity is clear: Batchewana First Nation territory lies adjacent to the territory of Garden River First Nation, and Sugar Island (and its islets) lies directly south, adjacent to Neebish Island. This area, composed of unceded Indian territory, represents perhaps the last remaining vestige of the original Indian territory, the original nation both the United States and Great Britain recognized as one element of the treaty that ended the Revolutionary War, the Treaty of Ghent, as well as the Royal Proclamation, the Northwest Ordinance, and other treaties and documents.

All of these treaties and documents point to one indisputable fact: the border that runs through the St Mary's River is nothing more than a convenient fiction the United States and Canada agreed to but never enacted by legislation and, as such, it can no longer be recognized by any thinking people as a border that bisects and disrupts the territorial integrity of the Indigenous North American nation of Bawating. Consequently, Bawating may be the ultimate point beyond the border – the exception that breaks the rule and leaves the border a fiction.

NOTES

1 The first Bull, the *Dum Diversas* (1452), set out the "authority" for enslavement, etc.; the second, the *Romanus Pontifex* (1455), granted certain rights to Portugal; the third, the *Inter Caetera* (1493), granted certain rights to Spain.
2 The term was first used by Garden River First Nation scholar Karl Hele (2008).
3 I serve as university liaison for the Anishnaabeg Joint Commission. This information comes from my personal involvement with these issues.

4 I served on the planning committee for this summit. The information included here is from my personal involvement. See also http://www.csibi.org.

REFERENCES

Barnes, J.K. 1867. *Circular, No. 2*. Washington, DC: Surgeon General's Office, War Department. http://www.flickr.com/photos/medicalmuseum/5033582606.

Batchewana First Nation, Garden River First Nation, and United States Customs and Border Protection. "Memorandum of Agreement between Batchewana First Nation, Garden River First Nation, and United States Customs and Border Protection." Undated draft copy circa 6 July 2010.

Bellfy, Phil. 2011. *Three Fires Unity: The Anishnaabeg of the Lake Huron Borderlands*. Lincoln: University of Nebraska Press.

Google Earth. Image copyright 2012 GeoEye.

Hele, Karl S., ed. 2008. *Lines Drawn upon the Water: First Nation and the Great Lakes Borders and Borderlands*. Waterloo, Ontario: Wilfrid Laurier University Press.

Hollinger, R. Eric, Gregory Anderson, Christopher Dudar, and Sarah Feinstein. 2009. "Inventory and Assessment of Human Remains and Funerary Objects from Ontario, Canada, Potentially Affiliated with the Sault Ste Marie Chippewa and Bay Mills Indian Community in the National Museum of Natural History, Smithsonian Institution." Unpublished manuscript. Washington, DC: Repatriation Office, National Museum of Natural History, Smithsonian Institution.

McNab, David T. 2004. "Borders of Water and Fire: Islands as Sacred Places and as Meeting Grounds." In *Aboriginal Cultural Landscapes*, edited by Jill Oakes and Rick Riewe, 35–46. Winnipeg: Aboriginal Issues Press.

Moore, John Bassett. 1898. *History and Digest of the International Arbitrations to which the United States has been a Party*. United States: US Government Printing Office.

Reuben, Thwaites, ed. 1883. "Saint Lusson's Process Verbal." *Collections of the State Historical Society of Wisconsin* 11: 26–9.

Sault Tribe of Chippewa Indians. 2012a. "Government." http://www.saulttribe.com/government.

– 2012b. "History." http://www.saulttribe.com/history-a-culture.

Struck, Doug. 2005. "Canada Fights Myth it was 9/11 Conduit." *Washington Post*, 9 April.

United Nations. 1945. Charter of the United Nations. June 26.

Conclusion: Beyond the Paradoxes of the Border

KYLE CONWAY AND TIMOTHY PASCH

We opened this book with the observation that the border serves a paradoxical function, suturing and separating two countries. We then considered this paradox by examining the mediated border, the political border, and the native border, in each instance parsing out the multiple valences of terms such as *mediated*, *political*, and *native*. We also noted that scholars of the borderlands have long drawn on the discipline of history, which has provided productive tools for thinking about regionality or for comparing and contrasting related phenomena on either side of the border. Despite the richness of this research, however, the paradigms history provides did not address the concerns raised by the participants in the conference out of which this book grew. Prompted by those participants' interventions, we wanted to address the border as a lived experience through the lenses of media, politics, and indigeneity. What has our approach yielded? Or, more to the point, what does this collection of essays yield when taken as a whole?

In the last chapter, Phil Bellfy's challenge to the border's very existence works to cast the other essays in this book in a new light. It asks us to think "beyond the border" – not only geographically, but also conceptually. It asks us to think about the nature of the tensions that shape this paradoxical line. The border of the first essays is altered by the final essays, and that alteration is a productive point of departure.

Consider, for instance, the comments of the anonymous US State Department official quoted in Serra Tinic's essay. This official described the Canada-US border as a "piece of real estate, which most Americans associate with snow blowing back and forth across an imaginary line" (10). Bellfy's challenge shifts our attention away from the piece of real estate to the imaginary line, with an emphasis on "imaginary." The line may be more ontologically fragile than the official's comments suggest, and its fragility must be compensated for through the processes of enforcement. The chapters that follow Tinic's consider one aspect of that enforcement, namely the discursive work that must be performed by governments, by people living on the border, by non-governmental organizations, and by many others to produce and maintain the different senses of national identity that distinguish Americans and Canadians who live in the Great Plains and Prairies.

In this conclusion, we would like to return to the Great Plains and Prairies. The questions Bellfy raises are relevant for the centre of the continent, and their answers reveal not only what the Great Plains and Prairies have in common with other regions, but also what makes them distinct. Bellfy asks several questions, including whether the creation of the border (not to mention its enforcement) represents a legitimate exercise of force. He additionally reflects on how the border has moved as a consequence of the history of treaties between the United States and Great Britain (before 1867) or Canada (after 1867). Finally, he asks about the effects of the vicissitudes of nature on where exactly the border lies.

To demonstrate the parallels between the Great Lakes region and the Great Plains and Prairies, we present answers to Bellfy's questions from two distinct perspectives. The first is historical, an examination of a stretch of border falling within the Great Plains, from Lake of the Woods (in northern Minnesota) to Pembina (in far northeast North Dakota, just south of the border with Canada in the Red River Valley). The second is contemporary, an examination of the border in an era of unmanned aerial drones for patrols and concerns about cyberwarfare. We conclude by describing the ramifications of the border's paradoxical function.

A BRIEF HISTORY OF THE FORTY-NINTH PARALLEL FROM LAKE OF THE WOODS TO PEMBINA

A number of points of continuity connect the Great Lakes region Bellfy describes and the section of the border connecting the Lake of the Woods to Pembina. They are geographically close: Lake Superior and Lake of the Woods are about two hundred miles apart. The same seventeenth- and eighteenth-century French *coureurs des bois* also initially explored them. Not only that, but some of the very same treaties established borders in both regions, in particular the Jay Treaty, which is central to Bellfy's claim about the illegitimacy of the border in the area of Sault Ste Marie.

There are important differences, too. The border between Lake of the Woods and Pembina was established at a different time than the stretch of border with which Bellfy is concerned. Although the treaties affecting it date back to the time of the US Revolutionary War, it was surveyed after the US Civil War, when many Americans were in an expansionist mood and when Canada, in the years that led up to and immediately followed Confederation in 1867, wanted to prove its resolve as a young country. The geography is also different: instead of a body of water with fluctuating levels, the area around Pembina in particular is exceptionally flat and subject to a different set of challenges.

Two issues arise when we consider this border. The first concerns its establishment in the first place: what negotiations led to its being where it is? The second concerns the process of surveying once the boundaries were established: how did the United States and Great Britain (later Canada) measure and mark the boundaries they had established? With respect to the first question, the relevant treaties were influenced not only by the political relationships between the United States and Great Britain/Canada but also by the refinement of surveying tools, which provided negotiators with increasing levels of precision. The first attempt to establish the boundaries of the United States dates back to the Revolutionary War itself. On 23 February 1779, a committee of Congress, whose task was to formulate the conditions of peace, recommended that the new country's boundaries be as follows: "Northerly by the ancient limits of Canada, as contended for by Great Britain, running from Nova Scotia, south-westerly, west, and north-westerly, to Lake Nepissing [near present-day North Bay, Ontario], thence a west line to the Mississippi ... and westerly by the river Mississippi" (quoted in Winchell 1898, 188). Congress considered the recommendations, and in the years that followed the new government negotiated with Great Britain to establish the new country's borders. The first mention of Lake of the Woods came in November 1782, when Richard Oswald, representing Britain, presented John Adams, Benjamin Franklin, and John Jay with a proposal that would establish the border "through the middle of ... Long Lake, and the water communication between it and the Lake of the Woods, to the said Lake of the Woods; thence through the said Lake to the most north-western point thereof; and from thence on a due western course to the river Mississippi" (quoted in Winchell 1898, 190). This same language figured in the Treaty of Paris in 1783, which marked the end of the Revolutionary War.

This language had two flaws, however. The first was that it was not entirely clear to later surveyors which lake was Long Lake (Lass 1980, 37). The second was more important: a line running due west from the Lake of the Woods would never reach the Mississippi River because the river did not (and does not) run that far north. The Jay Treaty addressed the growing suspicion that this was the case in its fourth article. The treaty stated that Britain and the United States should "[make] a joint survey" to determine whether such a line was possible, and if it was not, "the two parties will thereupon proceed, by amicable negotiation, to regulate the boundary line in that quarter" (quoted in Winchell 1898, 194). Although the survey was not made, negotiations did take place, and by 1800, astronomers had established that the northwestern corner of Lake of the Woods was farther north than the headwaters of the Mississippi River.

The first mention of the forty-ninth parallel came in a series of negotiations in 1807, although those negotiations broke down. The Treaty of Ghent in 1814, which marked the end of the War of 1812, spoke of the need to "fix and determine, according to the true intent of" the Treaty of Paris, flawed assumptions and all, "the part of the boundary between the dominions of the two Powers" of Great Britain and the United States (quoted in Winchell 1898, 201). It also provided for arbitration by "a friendly sovereign of State" in case the surveyors disagreed. The two countries' respective commissions did in fact present contradictory reports, but no arbitration took place. Great Britain and the United States revisited the issue in 1818, when they reached a convention whose second article stated that the border should fall along a line drawn due north or south from the northwest corner of Lake of the Woods to intersect with the forty-ninth parallel, and from there follow the parallel itself. It was the first time that the border was defined on paper, but even this convention did not settle the matter. The two countries revisited the question of the border again in 1842 with the Webster-Ashburton Treaty, which Bellfy also discusses.

Bellfy asks what the border looks like from the perspective of North America's native communities and, with respect to the border in the Great Lakes, judges it to be illegitimate. If we ask the same question here, it is hard not to arrive at a similar conclusion. The apparent illegitimacy of the border is two-fold. First, the treaties described here were all between European powers and their former colonies and were negotiated largely in Europe. First Nations were actively excluded and the imposed borders followed a European logic. The incommensurability between the native perspective and that of the European colonizers is clear in the name that First Nations had for the border: it was the Medicine Line, so-called because it "[had] 'strong medicine' since it seemed to have the power to stop the pursuing US Cavalry in its tracks" (Rees 2007, 5).

The border's illegitimacy was not merely a function of this exclusion. Instead, it resulted from the active suppression of First Nations as part of the project of colonial expansion, as the events leading up to the definitive survey of 1872 make clear. From Canada's perspective, the need to mark the border in a definitive way grew more urgent during the US Civil War. One concern was that because of Britain's support of the south, the United States might invade Canada "to compensate for possible territorial losses in the south. Others were convinced that, should the Union win, invasion would still occur because southern opposition to northern expansion would be removed" (Lass 1980, 77). Such fears were exacerbated by the introduction in the US Congress in 1866 of "An act for the admission of the states of Nova Scotia, New Brunswick, Canada East, and Canada West, and for the organization of the territories of Selkirk, Saskatchewan, and Columbia," which would have "provided that the eastern Canadian provinces would enter the United States as states, and the western regions would become territories" (Lass 1980, 77). Although the act had little real support, it was one factor that precipitated the process of Confederation in 1867.

Active suppression of First Nations took place in the years that immediately followed Confederation. The newly formed Canadian government, concerned about the prospect of annexation by the United States and eager to demonstrate its authority, moved to purchase Rupert's Land (which included present-day Manitoba) from the Hudson's Bay Company. It sent surveyors to the Red River Valley, where they met resistance from the Métis, led by Louis Riel, who saw the surveyors' arrival as a threat to their way of life. In October and November 1869, Riel led a group of Métis and occupied Fort Garry (in present-day Winnipeg), establishing a provisional government. In response, the Canadian government proclaimed the Manitoba Act in July 1870 and sent British and Canadian troops. Riel had to flee to Minnesota. Thus, not only were First Nations not included in negotiations about the border, the border was imposed through military force against them.

The fact that the border existed on paper, at least as of 1818, did not mean, however, that it was easily identifiable to people on the ground. This is the second point where its existence – its ontological status – proved a challenge. Before the survey of 1872, at least five markers purported to represent the boundary. The first was left in 1823 by US Army Major Stephen H. Long, and by virtue of being the first, it became the de facto border, even if later surveyors disputed its location. In 1857, for instance, British Captain John Palliser determined the forty-ninth parallel to be 370 feet north of Long's marker. Three years later, residents of Pembina, "offended by a whisky peddler who had set up shop just inside British territory," placed another marker a mile north of Long's marker (thus bringing the whisky peddler into US jurisdiction), although they did so "probably without benefit of astronomical observations" (Lass 1980, 79), and their marker was ignored by later surveyors. In 1869, Canadian Lieutenant-Colonel John S. Dennis identified yet another point as the forty-ninth parallel, this one about 200 feet north of Long's marker, and a year later, in the lead-up to the construction of Fort Pembina, US Captain David P. Heap concluded that the forty-ninth parallel was nearly a mile north of Long's marker. This would have placed a Hudson's Bay trading post within US territory, which would have created problems for US customs officers, who would have had to impose tariffs on all the goods contained there (Lass 1980, 79–80).

The problems raised by having multiple "official" makers were compounded by the vicissitudes of nature and the challenges of determining latitude in a region as flat as the Red River Valley. As Judy Larmour (2005, 10) explains:

> The forty-ninth parallel is not a straight line on the earth's surface, but actually a curved line, angling northward as the viewer looks west. The North American Boundary Commission intended to locate the forty-ninth parallel by astronomical

observations twenty miles apart. Agreement on how to survey the boundary became a major hurdle. The Americans proposed a straight line between each monument, but [chief British surveyor Donald] Cameron and the British commission quickly pointed out that Canada would have lost about two hundred acres every twenty miles! Canada successfully proposed a curved-line border following the parallel of natural latitude between each monument.

At the same time, the precision of their measurements was limited "from some influence, supposed to be due to irregularity of direction in the action of gravitation upon the levels of astronomical instruments" (Cameron, quoted in Thomson 1968, 213). Complicating their task further were the simple hazards of working outdoors in the Great Plains and Prairies, including mosquitoes, flies, snowstorms, grass fires, and even a lack of firewood for cooking (Thomson 1968, 213–14).

What this historical analysis reveals is that Bellfy's broad critiques are applicable in the Great Plains and Prairies, but they differ in substance. Instead of Europeans in the new world reading papal bulls at the continent's indigenous peoples, representatives of North American governments – still European in their modes of thought, if not in their political loyalties – sent armies to clear the way for colonial expansion. Instead of fluctuating water levels causing shifts in the "lines drawn upon the water," there was the challenge of surveying a land so flat that the conventional wisdom (according to which the forty-ninth parallel was a straight line) distorted the border itself.

But Bellfy's analysis is not only historical; it is also contemporary. Ours is, too. In particular, one of the more controversial and potentially polarizing issues related to the enforcement of borderlands in the Great Plains and Prairies is not situated primarily on land at all, but rather in the air.

CONTEMPORARY BORDER ISSUES OF THE GREAT PLAINS AND PRAIRIES: UNMANNED AERIAL SYSTEMS

On 4 February 2011, US President Barack Obama and Canadian Prime Minister Stephen Harper announced a joint initiative, called *Beyond the Border: A Shared Vision for Perimeter Security and Economic Competitiveness* (United States and Canada 2011), designed to facilitate increasingly synchronous approaches to border security.[1] The document explaining the initiative begins by emphasizing strong commonalities and links between the two countries: "The United States and Canada are staunch allies, vital economic partners, and steadfast friends. We share common values, deep links among our citizens, and deeply rooted ties. The extensive mobility of people, goods, capital, and information between our two countries has helped ensure that our societies remain open, democratic, and prosperous" (United States and Canada 2011, i). It then describes how a joint, perimeter-focused approach to security will catalyze economic prosperity by enhancing and facilitating the cross-border flow of goods, addressing threats, and "ensuring the safety, security, and resilience" of both nations (1). The plan identifies four distinct areas of cooperation, namely trade facilitation, economic growth, cross-border law enforcement, and critical infrastructure and cybersecurity.

Perimeter security has particular relevance to the borders in the Great Plains and Prairies, since one important initiative involves unmanned aerial surveillance aircraft deployed from Grand Forks, North Dakota, for border security and patrol. These unmanned vehicles, popularly called drones, reflect modern-day borderland concerns and tensions addressed by Bellfy and others, in that they offer the capacity to observe, at great distances, activities that take place either at checkpoints or at unsanctioned crossings. They are one piece in the puzzle, so to speak. On 20 April 2012, Royal Canadian Mounted Police and US Border Patrol Agents gathered at Altona, Manitoba, for the dedication of a new, shared border intelligence centre. This centre, the Red River Integrated Border Enforcement Team (IBET) joint intelligence office, serves as a facility where US and Canadian law enforcement agencies can share resources and data on border security issues. In recent years, a variety of IBET intelligence offices have come into being, with the Red River, Superior, and Prairie centres covering parts of North Dakota-Manitoba, Minnesota-Ontario, and Montana-Saskatchewan, respectively.

Although the idea of joint border intelligence operations has existed for some time, the events of 9/11 accelerated the development of these shared initiatives, with sophisticated observation technologies now contributing a critical component of border intelligence on the American and, increasingly, Canadian sides of the border. The boundaries that divide the Great Plains and Prairies are special: they cover thousands of miles of sparsely inhabited territory that has been poorly patrolled and is frequently inaccessible by regular transport. For these reasons, a number of technologies have been used to protect this border remotely, including the Integrated Surveillance Intelligence Systems, or ISIS (buried seismic and magnetic detectors), unmanned aerial systems (drones), and remote video surveillance systems (remote cameras). In the case of the vast spaces of the Great Plains and Prairies, the unmanned aerial systems gain the most attention and generate the most controversy.

On 28 April 2012, US Homeland Security Secretary Janet Napolitano stated that the use of unmanned aerial aircraft for border surveillance between the United States and Canada had expanded from North Dakota to eastern Washington. In a related congressional research service report, a military policy analysis on the effectiveness of drone patrols described the particular complexities of the border and the challenges inherent in policing it:

> Border Security has long been recognized as a priority by the Congress. The northern border separating the mainland United States and Canada is 4,121 miles long and consists of 430 official and unofficial ports of entry. The expansive nature and the possibility of entry through unpopulated regions make the border difficult to patrol.
>
> One potential benefit of UAVs (Unmanned Aerial Vehicles) is that they could fill a gap in current border surveillance. In particular, technical capabilities of UAVs could improve coverage along remote sections of the US borders. Electro-Optical (E-O) identification technology is advanced enough that it can identify a potentially hostile target the size of a milk carton at an altitude of 60,000 feet. (Blazakis 2004, 4)

In addition, the type of UAVs launched from Grand Forks can stay aloft for extended lengths of time, some for more than twenty hours, which makes continuous observation possible over the Great Plains and Prairies, and indeed across the entire border.

Although more prevalent in the popular media than ever before, these UAVs are not entirely new: the same unmanned aircraft that now patrol the Canada-US border have been deployed along the Mexico-US border since 2007. The fact that the aircraft patrolling the "open wound" of the Mexico-US border (in Gloria Anzaldúa's words, as described in Feghali's chapter) now patrol what was long characterized as the world's longest undefended border has caused great concern. In particular, the dramatic increase in surveillance technology provides a stark contrast for those who recall Canada-US border crossings before 9/11. Despite the assertion that the "United States and Canada are staunch allies, vital economic partners, and steadfast friends," people who remember the border before 9/11 now ask whether the increasing militarization reflects a different type of relationship. For instance, domestic drone use was the topic of a December 2011 special report by the American Civil Liberties Union, which envisioned a dystopian future of ubiquitous unmanned aerial surveillance designed to record all movement (across the border or within a nation) with sophisticated analysis to predict potentially suspicious behaviour and make pre-emptive arrests (ACLU 2011). This transposition of technology has significance that goes potentially deeper than the merely technical. Research at the Border Policy Research Institute at Western Washington University suggests: "Actual definition[s] of security and insecurity at the border [become] articulated in military terms by the military" (Muller 2008, 13). For residents on any side of a border, there is something unsettling about being watched from above, especially when drone observation "abstracts people from contexts, thereby reducing variation, difference, and noise that may impede action or introduce moral ambiguity ... When these mechanisms and logics of surveillance are imported to non-combat settings, such as borderzones and civilian territories, they may in turn further the violent dehumanization and non-differentiation of people while expanding the scope of who could be included in the drone's gaze" (Wall and Monahan 2011, 239–43).

Parallels also exist between the introduction of unmanned aerial vehicles for border patrol between the United States and Canada and Tinic's analysis of the Canadian television show *The Border*. Tinic writes: "The central narrative of the series focuses on the Canadian agents' consistent 'education' of the archetypical American 'cowboy' officer in matters of tolerance to cultural difference and respect for jurisdictional differences in matters ranging from native land claims to attitudes about terrorism" (36). Although the domestic audience for the series was small, Tinic explains that it was closely monitored by the US State Department, which raised concerns about the use of Canadian tax dollars to create a show that reinforced negative stereotypes in the United States. Now, however, the images Canadians see are not on a fictional program but on the news, such as on the 7 April 2011 edition of CBC's the *National*, which featured a story about drones launched from Grand Forks. The story opened with footage of US drones bombing Iraq and Afghanistan and then described how similar drones are now deployed along the Canada-US border. Especially alarming was the idea that such drones could now fly right along the border, while before they had to stay at least sixteen kilometres away. This was a cause for concern for Canadians worried about privacy and American motivations for coming so close to the border. Along with the near ubiquity of drone bombing reports from the Middle East, it was also the source of unfounded concerns that these drones are capable of delivering militarized attacks, concerns that were perhaps due, in part, to the influence of television programs such as *The Border*.

Although much popular discussion of unmanned aircraft may be alarmist, these vehicles are not only used for patrolling the border: they also have applications particular to life in the Great Plains and Prairies. They have been used during emergencies for mapping flooded regions of the Red River Valley in North Dakota and Minnesota, for example, and for search-and-rescue operations. The heat-sensitive cameras in the aircraft (able to detect facial features and the heat from a cigarette many kilometres away) are also effective for mapping water movements and boundaries. Thus their potential to serve the repressive state apparatus (as Moore might point out) or the forces of neocolonial expansion (as Bellfy might point out) is only one side of the coin.

It should be noted, however, that the United States is not the only country that uses drones. Taking cues from its southern neighbour, Canada is considering the use of drones to address concerns with Arctic sovereignty and the use (or potential misuse) of the Northwest passage. A proposal for an Arctic Hawk (a modified version of the RQ-4B Global Hawk UAV currently used on the Canada-US border) was described by Canadian Defense Minister Peter Mackay in June 2012 as a potential option for Canadian Arctic sovereignty (Leithen 2012).

COMBINED DIGITAL INFRASTRUCTURE AND CYBERSECURITY

Unmanned aerial surveillance is not the only aspect of increased militarization at (and above) the Canada-US border. Another aspect

Figure 10.1 A UAV deployed for border patrol from the Grand Forks Air Force Base in North Dakota that will fly over the Great Plains as far west as Spokane, Washington (*The National*, 7 April 2011)

of the integrated, cross-border law-enforcement mentioned in the *Beyond the Border* document is the initiative for "enhanc[ing] our already strong bilateral cybersecurity cooperation to ... increase both countries' ability to respond jointly and effectively to cyber incidents" (United States and Canada 2011, 23).

Cyberattacks and cybercrime are an increasing area of concern regarding threats to the national security of Canada and the United States. Cybercriminals have the ability to cross the border virtually at will, and so evade border security even more adeptly than those who attempt to physically evade the gaze of aerial unmanned vehicles. The *Beyond the Border* initiative identifies a clear need for the United States and Canada to respond in tandem to counter cyberthreats.

Perhaps more so than with aerial surveillance, where most UAVs originate and are controlled from the United States, Canada may be able to contribute more equally to the joint venture of defending against electronic threats. To do so, however, communities on both sides of the border share a great need for rapid connectivity. For this reason, especially in the Great Plains and Prairies, the completion of the final leg of the Northern Tier Data Network, which would connect North Dakota and Manitoba for high-speed research and data transfer, would be a significant step forward.

Since the IBET joint border intelligence organizations are still new, joint cybersecurity efforts are an area to observe carefully. International borders and boundary lines are seldom respected in cyberspace. The arbitrariness that Bellfy describes of the "lines drawn" between countries takes on a new dimension and highlights a need for research into the roles of policing the border, whether with technology or with people. This type of enforcement works in tandem with the discursive work described in the preceding chapters. As Moore reminds us, the repressive state apparatus, in the form of the military, of border security agencies, and so on, stands ready to step in when the ideological state apparatuses fail. This is one area of research that moves beyond the border conceptually, and we predict that questions of the mutual dependence of policing and discursive work will grow in importance as the emphasis on border security increases.

As important as security is, we would be remiss if we did not reflect on other, more intangible movements across the border. Even while it becomes ever more intimidating, secure, and even militarized, other relationships flourish. At the Métis festival in the International Peace Garden mentioned in this book's introduction, for example, Franco-Manitoban fiddlers played jigs with Métis Nation drummers from Montana and the Dakotas. Franco-Manitoban singer Daniel Lavoie sang of days on the prairies in his song "Jours de plaine" at the 125th anniversary tribute of Louis Riel's death. It was clear that the deep significance of life in the prairies resonated with people on both the Canadian and US sides of the border. The Michif language still spoken in the Turtle Mountain Band of Chippewa resonates with students at the Université de St-Boniface in Manitoba. Each spring, residents of the Red River valley from Fargo to Winnipeg lay down sandbags to protect their homes from the rising waters. The courtesy and helpfulness and sense of beauty on both sides of the border transcend it, and prompt us to rethink it as an interstitial place where identities can re-emerge. After each crossing, we move forward with a sense of relief, and a burgeoning excitement at discoveries to be made on the other side in these strange and new cultural atmospheres, intangibly yet palpably different from our own.

Which, then, is it? The permeable border, or the immovable one? The answer, it seems, is a little of both. As the lines become stronger, and as surveillance technologies grow in sophistication, we must take care to guard against a situation where the border becomes so challenging to cross that it becomes prohibitive even to make the effort. The risk is that not only will the "bad" contact be arrested, but that the "good" – and deeply desirable – contact, too, may be arrested or prevented altogether. What Bellfy, Feghali, and Miner help us to understand is that the crossing in and of itself, especially (but not only) for First Nations, can be a restorative experience for both sides and should be promoted and encouraged.

There are other questions to ask as well. Two of the presenters at the conference out of which this book grew occupied senior positions within the Canada Border Services Agency and the United States Border Patrol. Terrorism and drug contraband were topics mentioned continually as among the most urgent issues on both sides of the border. If we are to attempt to find a compromise between an increasingly militaristic border and what used to be the world's longest undefended border, we need to look not only outward with an eye to preventing entry to those who might want to breach these lines, but also inward at the policies and laws that provide economic and political incentives that cause them to try to do so. To begin, we must ask fundamental questions about how to find a balance between border security and the lived realities of people who cross the border, so that the force required to maintain border security does not become overtly, or even prohibitively, oppressive. It is incumbent to realize, not only in the Great Plains and the Prairies but all across the Canada-US (and Mexico-US) border, that many of the issues we currently face are symptoms or direct results of foreign policy decisions, and that increasing security at the border will not in and of itself address these root issues.

To sum up, this work is, ultimately, a compromise. With narratives that move from an inviolate, immovable border, traverse the ideological gradient in degrees, and finish at a conceptual border that is nothing if not porous (in contrast to border as we experience it now), we want to be clear that we are not suggesting that the goal is to completely re-imagine the border, or even to change it significantly in the short term. Rather, we propose a re-examination of the beliefs and causes in each country that seem to indicate the necessity for ever-increasing border security since 9/11. We propose that these essays can help ask the hard questions of how to make space for the paradoxes and contradictions of life along the border and reduce the need for the repressive forms of security that no one – on either side of the border – wants to see awaiting them when they begin their own crossing.

NOTE

1 It was a matter of coincidence that when we submitted the original manuscript of this book a year before, we had titled it *Beyond the Border*. The title has proven fortuitous.

REFERENCES

ACLU (American Civil Liberties Union). 2011. "Protecting Privacy from Aerial Surveillance: Recommendations for Government Use of Drone Aircraft." 15 December. http://www.aclu.org/technology-and-liberty/report-protecting-privacy-aerial-surveillance-recommendations-government-use.

Blazakis, Jason. 2004. *Border Security and Unmanned Aerial Vehicles.* CRS Report for Congress, 2 January. Washington, DC: Library of Congress Congressional Research Service. http://epic.org/privacy/surveillance/spotlight/0805/rsjb.pdf.

Larmour, Judy. 2005. *Laying Down the Lines: A History of Land Surveying in Alberta*. Victoria, BC: Brindle and Glass.

Lass, William E. 1980. *Minnesota's Boundary with Canada: Its Evolution Since 1783*. St Paul: Minnesota Historical Society Press.

Leithen, Francis. 2012. "Canada Outlines Arctic UAV Requirement." 5 June. *Aviation Week*. http://www.aviationweek.com/Article.aspx?id=/article-xml/awx_06_04_2012_p0-464531.xml.

Muller, Benjamin J. 2008. *Governing through Risk at the Canada/US Border: Liberty, Security, Technology*. Bellingham, WA: Western Washington University Border Policy Research Institute.

Rees, Tony. 2007. *Arc of the Medicine Line: Mapping the World's Longest Undefended Border Across the Western Plains*. Lincoln: University of Nebraska Press.

Thomson, Don W. 1968. "The 49th Parallel." *Geographical Journal* 134 (2): 209–15.

United States and Canada. 2011. *United States-Canada Beyond the Border: A Shared Vision for Perimeter Security and Economic Competitiveness.*

http://www.whitehouse.gov/sites/default/files/us-canada_btb_action_plan3.pdf.

Valdes, Manuel. 2012. "Homeland Security Drones Patrol Washington-BC Border." *KOMO News*, 28 April. http://www.komonews.com/news/local/Homeland-Security-drones-patrol-Washington-BC-border--149356585.html.

Wall, Tyler, and Torin Monahan. 2011. "Surveillance and Violence from Afar: The Politics of Drones and Liminal Security-scapes." *Theoretical Criminology* 15 (3): 239–54.

Winchell, Alexander N. 1898. "Minnesota's Northern Boundary." In *Collections of the Minnesota Historical Society* 8: 185–212. St Paul: The Society. http://books.google.com/ebooks?id=5YtuAAAAMAAJ.

Contributors

PHIL BELLFY is a member of the White Earth Band of Minnesota Chippewa and professor emeritus of American Indians Studies at Michigan State University. He is the founder and co-director of the Center for the Study of Indigenous Border Issues and the editor and publisher of its Ziibi Press. His most recent book, *Three Fires Unity: The Anishnaabeg of the Lake Huron Borderlands* (2011), won the University of Nebraska Press "North American Indian Prose Award" for 2010. And he really can see Canada to the North and the East from his house in rural Sault Ste Marie, in Michigan's Upper Peninsula.

KYLE CONWAY is an assistant professor of communication at the University of North Dakota. He works in the fields of media studies and translation studies, paying special attention to acts of negotiation that take place at points of contact in literal and metaphorical border-zones. He is the author of *Everyone Says No: Public Service Broadcasting and the Failure of Translation* (2011).

CHRISTOPHER CWYNAR is a PhD student in media and cultural studies in the communication arts department at the University of Wisconsin-Madison. His research interests include nations and nationalism, public service broadcasting, and digital convergence. His dissertation addresses the CBC's engagement with the Internet from the mid-1990s to the present. He has presented his work at a number of conferences, including the Canadian Communications Association and the Society for Cinema and Media Studies.

BRANDON DIMMEL is a PhD candidate in the department of history at Western University in London, Ontario. His research focuses on cross-border conflict and cooperation at the local level during the First World War. His dissertation is entitled "Outside Influences: Great War Experiences along the Canada-US Border." He has previously published articles based on his research in the *American Review of Canadian Studies* and the *Journal of Borderlands Studies*. He currently teaches Canadian history at the University of Windsor.

ZALFA FEGHALI earned a PhD on queer citizenship and border studies at the University of Nottingham. Her next project maps a literary history of citizenship in the North American context.

JOSHUA D. MINER is a doctoral candidate and instructor in American literary studies and indigenous studies at the University of Iowa, where is he completing his dissertation on health in twenty-first century native literatures. He holds a BA in creative writing and MA in literature and linguistics from the University of North Texas. In his work, Miner writes from both his youth in southern Red River country and the transnational experiences of his adulthood. His most recent publications include poetry and stories in *Danse Macabre* and *Broad River Review*. He was a finalist for the 2012 Trio Award as well as the 2012 Rash Award in Poetry.

PAUL MOORE is associate professor of communication and culture at Ryerson University in Toronto. His overview of the first year of cinema coast-to-coast in Canada recently appeared in the *Canadian Journal of Film Studies*; his other studies of early cinema in North America were published as chapters in *Explorations in New Cinema History* and *A Companion to Early Cinema*. His forthcoming book with Sandra Gabriele, *The Sunday Paper*, explores the intermedial circulation of illustrated weekend newspapers in North America.

MICHELLE MORRIS is a PhD student in social and ecological sustainability at the University of Waterloo and a member of the multi-university Water Policy and Governance Group. Her research interests include transboundary water governance, collaborative approaches to water governance, water security, and water allocation. She has MA and BA (honours) degrees in political science from the University of Alberta and the University of Lethbridge, respectively.

TIMOTHY PASCH (PhD, communication, University of Washington 2008) is an assistant professor of communication at the University of North Dakota. Pasch is a dual US-Canadian citizen and fluent French and Japanese speaker, with some training in Inuktitut. His research interests focus on issues of cyberculture and the digital humanities, especially as they pertain to the circumpolar Arctic and interstitial borderlands. He leverages and integrates disparate creative technologies for the purposes of linguistic and cultural preservation, learning enhancement, digital human rights, and augmented collaboration environments.

PAUL R. SANDO is an associate professor of geography at Minnesota State University Moorhead. Though his education has taken him to such far-flung places as Indiana, Texas, Mexico, Latvia, and the Baltic region, he was born in and is a long-time resident of the Red River Valley. His interest in border issues with Canada stems from an interest in travel, education, and life along the border. His interest in flooding and water issues comes from living in two flood-prone cities in the region, and a longtime association with border agriculture. He currently lives with his family a mere five blocks from the Red River of the North.

SERRA TINIC is an associate professor of sociology at the University of Alberta. Her research focuses on the global television industry, with special attention to Canada's place in that industry. She is the author of *On Location: Canada's Television Industry in a Global Market* (University of Toronto Press 2005).

Index

9/11 (11 September 2001): and First Nations, 154, 158–9, 207–8, 212, 215–16, 219; and security, 20, 158–9, 206–8, 230, 232, 236; and television, 30, 35–6, 41, 44, 54, 59; and "world's longest undefended border," 3, 163, 206

Adorno, Theodor, 9
Alameda Dam, 138
Alaska, 208, 212
Alberta, 17, 77, 85, 185; and Alberta Agriculture and Rural Development, 116, 118, 124–6, 128; and Alberta Environment, 116, 118, 124–8; and Alberta Wilderness Association, 127–8; irrigation districts, 125–6; and Southern Alberta Group for the Environment, 127–8. *See also* International Joint Commission; Milk River; and Montana
Alexander VI (Pope), 200
Althusser, Louis, 12–14, 73–4
American Idol (television show), 55

Anishinaabe nation, 199–221; in literature, 164, 172–3, 177, 181–2, 187. *See also* Chippewa nation; Erdrich, Louise; Ojibwe nation; Vizenor, Gerald
Anzaldúa, Gloria, 3, 19–20, 153, 155, 159–65, 231
Asbeck, Richard, 94, 102–3, 105
Assiniboine nation, 129n3, 175
Assiniboine River, 17, 133, 137

Batchewana First Nation, 214, 216–20
Bawating (Lake Superior), 199–221
Bay Mills Indian Community, 214–19
Beaty, Bart, 44–5
Beyond the Border initiative, 229–32, 234, 237n1
Bierce, Ambrose, 153
Bismarck (North Dakota), 18
Blackfoot nation: and literature, 22, 170, 173, 175–6, 181, 185, 189; and riparian rights, 17, 113, 116–19, 121–3. *See also* King, Thomas

Blaine (Washington), 8, 14–15, 93–4, 96–110
borderland studies: comparative approach, 5; continentalist approach, 5; geographic approach, 4–6; historical approach, 5. *See also* Anzaldúa, Gloria; Hele, Karl; Konrad, Victor; Newman, David; Nicol, Heather; Paasi, Anssi
Border Patrol (United States), 158–9, 230, 236
Boundary Waters Treaty (BWT), 15, 119, 122
Brébeuf, Jean de, 181
British Columbia, 14, 75, 77–8, 84, 86–7, 227. *See also* White Rock
Brown, Edward, 101, 107
Bruyneel, Kevin, 174, 176, 178
Bryan, William Jennings, 101, 107
Byers, Michele, 45, 48, 50
Byrd, Jodi, 172, 179

Canadian Broadcasting Corporation (CBC), 10, 33–5, 37, 44–5, 56, 63; history of, 31, 48; and mandates, 11–12, 33, 37, 47; and news, 44, 233–4; values of, 12, 37, 41–2, 44, 47, 49–51. *See also Little Mosque on the Prairie*
Canadian Pacific Railway (CPR), 14, 73, 75–6, 80, 85–7
Champagne, Duane, 178, 180
Charland, Maurice, 44, 48
Chertoff, Michael, 21, 159, 166n3. *See also* Department of Homeland Security
Chicago (Illinois), 13, 74, 76–7, 80, 83, 86
Chicano political movement, 160–2

Chippewa nation, 19, 22–3, 199, 209, 213–18, 235; in literature, 173, 178, 187, 190. *See also* Anishinaabe nation; Ojibwe nation
Civil War (United States), 225, 227
Clean Air Act, 159
Clean Water Act, 159
Clinton, Hillary, 156
Coastal Zone Management Act, 159
Cocopah nation, 158
Conquest (1760), 201–2
Constitution Act (Canada), 123, 208
Cosgrove, John, 84–6
Cryderman, John, 13, 79–81, 84, 86
Custer, George Armstrong, 176
cybersecurity, 230, 233–5

Dakota (Bwaaneg) nation, 183
Da Vinci's Inquest (television show), 34–5
Degrassi (television franchise), 50, 65n10
Dennis, John S., 228
Department of Homeland Security (US), 13, 36, 231; and First Nations, 21, 154, 159, 219; and WikiLeaks, 10–11. *See also* Chertoff, Michael; unmanned aerial vehicles
Department of State (US), 6, 13; and First Nations, 18–19, 156; and WikiLeaks, 10–11, 18, 30, 37, 223, 232
Devils Lake (North Dakota), 17; and early cinema 81–2; and

flooding, 7, 17, 133–5, 140–3, 146–9
diphtheria, 99–100, 106
Dr Davison's Museum of Anatomy, 82–3, 86–7
Driskill, Qwo-Li, 180–1
Druick, Zoë, 52–3
Due South (television show), 40
Duluth (Minnesota), 81, 83
Dum Diversas (papal bull), 200, 220n1
Dymond, Greig, 45

earth-diver myth, 190
Edison, Thomas, 13, 71, 78–80, 86
Endangered Species Act, 21, 159
English Canadian identity, 9–11; and media, 29–37, 39–63, 95
environmental non-governmental organizations (ENGOs), 17, 116–18, 127–8
Erdrich, Louise, 22, 179; *The Last Report on the Miracles at Little No Horse*, 183; *The Plague of Doves*, 179, 181, 183–4, 187
erotic conversion, 173, 180–2, 193–4
Evans, Craig, 148

Fargo (North Dakota): and early cinema, 71, 76, 78, 80–3; and Red River, 13, 143, 145–6, 235
Fenneman, Nevin, 4
Fergus Falls (Minnesota), 78, 82
First Nations. *See specific nations*
First World War, 93, 95, 98, 109–10
Fort Garry (Manitoba), 71, 228
Freer, James S., 85
Friesen, Gerald, 73

Frozen River (film), 18–19
Frye, Northrop, 48–9
FUNdamentalist Films, 42

Garden River First Nation, 206–7, 214–20
Grand Forks (North Dakota), 6, 230–1, 233–4; and early cinema, 71, 76, 78, 80–3, 86; and Red River, 18, 143
Great Britain: and early cinema, 78, 85; and First Nations, 19–20, 156–7, 201–5, 208–10, 215; influence on Canadian culture, 39, 52, 62, 74; and the United States, 8, 20, 154, 158, 201–5, 208–10, 220, 224–7
Great Lakes, 4, 20, 22–3, 95, 224; and First Nations, 8, 199–220, 227; and literature, 181, 188
Great Northern Railway, 97, 100, 103, 110; and early cinema, 76, 81; and literature, 173, 177, 184
Gros Ventre (A'ani) nation, 175. *See also* Welch, James
Group of Seven, 95

Habermas, Jürgen, 6, 15–16
Haddock, Chris, 34–5
Hardie, Richard, 13, 74, 79, 81, 84–7
Harper, Stephen, 229–30
Heap, David P., 228
Hele, Karl, 4–5, 153, 165, 206
hemispheric studies, 154–5, 162, 165
Hendry, John, 97
Hill, James J., 75–6, 177
Historic Preservation Act, 21, 159

Hollywood, 29–30, 32–5
Horkheimer, Max, 9
Hudson Bay, 113, 136, 140, 142
Hudson's Bay Company, 20, 71, 75, 203, 227–8
Huron-Robinson Treaty (1850), 210, 215, 219
Hutcheon, Linda, 52

ideological state apparatus (ISA), 13–14, 73, 235
Indian Health Service, 185
Innis, Harold, 39, 74–5
Integrated Surveillance Intelligence Systems, 231
Inter Caetera (papal bull), 200, 220n1
International Joint Commission (IJC), 15, 100, 115; influence of, 17, 117–22, 125–28; and Order on Apportionment (1921), 17, 115, 117. *See also* Boundary Waters Treaty
International Peace Garden, 19, 235
Iroquois nation, 19–20, 156–7, 165
Isern, Thomas, 5
Islam, 12, 36, 42–3, 50–1, 58; and Christianity, 46–8, 51, 56, 59, 61; and fundamentalism, 54–5; and Islamophobia, 36, 41; and Orientalism, 44, 46, 51, 54–5, 58; and stereotypes, 11, 43, 46, 48–9, 53–9, 61–2

Jay Treaty. *See* Treaty of London (1794)
Jesuits, 182, 186–7, 193
Jones, Wesley, 101
Justice, Daniel Heath, 178, 193

Kainah nation, 17, 115–19, 121–4
Keary, John, 106–7
Kickapoo nation, 158
King, Joseph H.T., 216–17
King of Kensington (television show), 45–6
King, Thomas, 170–1, 173, 179, 182, 185, 189, 194; *Truth and Bright Water*, 182, 189–90
Klein, Kurt, 126–7
Konrad, Victor, 156, 162–3

Lake Agassiz, 142
Lake Darling Dam, 138
Lake of the Woods, 204, 224–6
Lake Sakakawea, 139
Lake Souris, 137
Lake Superior: First Nations, 22–3, 199, 203, 209, 224; transportation, 75, 84
Lake Winnipeg, 7, 16–17, 142
Little Mosque on the Prairie (television show), 7, 11–12, 36–7, 41–63; and regionalism, 49, 51. *See also* Islam
Lock, Helen, 21, 164
Long Lake, 226
Long, Stephen H., 228
Louis XIV (King), 200–1
Lower, Arthur, 95

Macdonald, John A., 95
McCarthy, William, 84–5
McMillan & Wife (television show), 61
McNab, David, 208
manifest destiny, 174, 176
Manitoba, 113, 235; and Devils Lake, 140, 142–3; and early film,

72, 74–5, 77–81, 84–7; and First Nations, 19, 187, 227–30; and Red River, 7, 13, 15, 17–18, 143; and Souris River, 137–8
Manning, Frank, 40, 47
Maple River, 145
Matheson, Sarah, 44, 46
Marshall, T.H., 157, 166, 168
Martz, Judy, 115
"medicine line," 165, 171, 227
mestizaje/métissage, 160–4. *See also* miscegenation; xenogamy
Métis, 19, 227–8, 235; and literature, 171, 181, 183, 187, 190
Mexico-US border, 3, 20–1, 231, 236; and First Nations, 157–9; as model, 8, 19, 153–9, 162, 166
Michigan: and early cinema, 82; and First Nations, 199, 209–10, 212–15
Migratory Bird Treaty Act, 159
Milk River, 7, 15–17, 113–29, 191, 193
Miller, Mary Jane, 12, 40
Minnesota, 224, 228, 230, 233; and early cinema, 72, 75–8, 81–3; and flooding, 143–4; and literature, 177, 187, 192
Minot (North Dakota), 137, 139–40. *See also* Souris River
Mi'kmaq nation, 157
miscegenation, 173, 181, 187–9. *See also mestizaje/métissage*; xenogamy
Mississippi River, 113, 193, 225–6
Missouri River, 7, 16, 113, 134–6, 139–40, 146
Mitchell, Chief Mike (Akwesasne), 208

Mohawk nation, 157
Monroe, Marilyn, 189
Montana, 230, 235; and early cinema, 76, 79; and literature, 173, 177, 181; and St Mary and Milk Rivers, 16, 113, 115–23, 125–6, 128. *See also* Alberta
Mooney, James, 182
Moorhead (Minnesota), 75, 143, 145. *See also* Red River
Mouse River. *See* Souris River
multiculturalism, 10–11, 41–2, 45–51, 54–5, 61

National Environmental Policy Act, 21, 159
National Historic Preservation Act, 159
Native American Grave Protection and Repatriation Act (NAGPRA), 159, 217
Nawaz, Zarqa, 36, 42–3, 46, 64n6. *See also Little Mosque on the Prairie*
Neebish Island, 205–6, 209–12, 220
negative identity, 29, 37
New Brunswick, 227
New France, 181
Newman, David, 5–6, 10, 14–16, 18, 23n1
Nicholas V (Pope), 200
Nicol, Heather, 156, 162–3
Nin, Anaïs, 184, 186
North Dakota, 15, 17, 19, 136; and early cinema, 79–82, 86; and literature, 173–4, 179, 184, 187; University of, 3, 184; and unmanned aerial vehicles, 231, 234; and State Water

Commission, 142. *See also* Fargo; Grand Forks; Minot; Pembina; Red River; Souris River
Northern Pacific Railway, 75–6, 83, 97
Northwest Area Water Supply Project (NAWS), 133, 135, 138–40, 146
Northwest Mounted Police, 73
Northwest Ordinance, 202, 205, 220
Northwest Territories, 14, 87, 208
Nova Scotia, 225, 227

Obama, Barack, 229–30
Ojibwe nation, 22–3, 183, 187, 199. *See also* Anishinaabe nation; Chippewa nation
Ontario, 31, 74, 84, 95, 225; and First Nations, 199, 208, 214–15, 218
Oswald, Richard, 225
Ottawa (Ontario), 10, 18, 95, 134
Outer Limits (television show), 32
Owens, Louis, 171–3

Paasi, Anssi, 5–6, 10, 14–16, 18, 23n1
Pacific Highway, 103
Palliser, John, 228
parody, 12, 42, 52–5, 59–60
Peace Arch, 93–4, 96, 108
Pembina (North Dakota): and border, 224–5, 228; and early cinema, 13, 71, 76, 80, 82
Pevere, Geoff, 45
Pick-Sloan project, 139, 146–7
Poindexter, Miles, 107
Pulitano, Elvira, 22, 155

Quebec, 46, 95, 208

Raboy, Marc, 31
Raff & Gammon (marketing firm), 78–80
Rafferty Dam, 138
Real ID Act, 20–1, 159, 166n2
Red River, 7; and border, 224, 227–8, 230, 233; and early cinema, 12–13, 71–87; and literature, 183, 187, 193; and water management, 17–18, 133–6, 140, 142–9
Regina (Saskatchewan), 36, 137
repressive state apparatus (RSA), 13–14, 73, 233, 235
reserves/reservations, 174–7, 182; Fort Belknap, 175, 177; White Earth, 187
Revolutionary War (United States), 202–3, 220, 225–6
Rice, Sir Cecil Spring, 101
Rick Mercer Report (television show), 35
Riel, Louis, 227–8, 235
Rocky (film), 59–60
Rocky Mountains, 4–5, 86, 124
Romanus Pontifex (papal bull), 200
Royal Proclamation (1763), 201, 220
Rukszto, Katarzyna, 52–3

Sadowski-Smith, Claudia, 156, 162, 165
Said, Edward: and travelling theory, 19–20, 154–5, 160, 162, 166; and Orientalism, 54–5
St Cloud (Minnesota), 78, 82
St Leonard Hotel (White Rock, BC), 15, 93–4, 96–110

Index

St Mary River (Great Plains/ Prairies), 7, 15–17, 113–29
St Mary's River (Great Lakes), 199, 203–6, 209–12, 214–16, 218, 220
St Paul (Minnesota), 75, 82, 97, 148
salmon fishing, 95, 97–100, 103, 105, 108–10
Sando, Todd, 146
Sands, Charles, 105
San Francisco (California), 76, 80, 83
Saskatchewan, 227, 230; and early cinema, 77, 85–6; and television, 11–12, 36, 42–3, 47; and South Saskatchewan River Basin, 16–17, 116, 124, 126, 128; and St Mary and Milk rivers, 113, 119; and Souris River, 17, 137–8
satire, 12, 42, 52–4, 62
Sault Ste Marie (Michigan/ Ontario), 199, 201, 207, 213–19
Schoolcraft, Henry Rowe, 209, 212–13
Secure Fence Act, 158
Semiahmoo Bay, 96–100, 104–5, 107–10
Sheets, J.W., 101–1, 105–7
Shepard, R. Bruce, 5
Sheyenne River, 140, 142, 146–7, 149
Sinclair, F.D., 99
Sitting Bull, 176
Smith, Goldwin, 95–6
Smith, John, 172
Smithsonian Institution, 216–19
Smythe, Dallas, 9, 30, 34
Souris River, 7, 17, 133–9, 145–9; dams, 138; International Souris Basin Commission, 139, 147

sovereignty, 5–6, 15, 21–2, 29; and Arctic, 208, 233; and First Nations, 156–7, 166, 170–94, 201–5, 216–20; and Quebec, 46
Spirit Lake Tribal Organization, 147
Spring-Rice, Cecil, 101
Stargate (television show), 32
Stewart, Margaret, 96–8
Street Legal (television show), 40
Sugar Island, 205–16, 220
Sullivan, Rebecca, 44–5
Surrey County (British Columbia), 96–110

Taylor, Charles, 9–10
Taylor, Lorne, 125
technological nationalism, 44
television: and drama, 32; and genre, 34, 40–2, 59–60, 62–3; and international joint ventures, 32; and runaway productions, 33; and situation comedy, 11–12, 45–51, 53, 60; and sketch comedy, 52
Texas, 14, 82–3, 158
The Border (television show), 10–11, 36–7, 41, 232
Three Affiliated Tribes, 147
Thrift, Henry T., 106–7
Tijuana River Research Reserve, 21, 159
Tohono O'odham nation, 157–8
Tolna Coulee, 142, 149
Toronto, 74, 95; and television, 12, 33, 42–3, 46, 48–9
transmotion, 177–9
Treaty of Detroit (1855), 215, 219
Treaty of Ghent (1814), 71, 158, 204–5, 209–11, 220, 226

Treaty of Guadalupe-Hidalgo
  (1848), 158
Treaty of London (1794): and border, 225–6; and First Nations, 20, 23, 154, 158, 203–5, 207–8, 213, 217
Treaty of Paris (1783), 203–4, 226
Treaty of St Petersburg (1825), 208
Treaty of Washington (1836), 158, 209–11, 215, 219
trickster, 21, 164, 183. *See also* Vizenor, Gerald
Turtle Mountain (Chippewa), 19, 22, 173, 178, 187, 235. *See also* Erdrich, Louise
two-spirit, 180–1

United Nations, 16, 219
United States Army Corps of Engineers, 143, 148
United States Geological Survey, 147
unmanned aerial vehicles (UAVs), 224, 229–4

value-added industries, 126–7
Vancouver (British Columbia), 33–4, 76–7, 83; and White Rock, 99–100, 104–6, 109
vaudeville, 77
Victoria (British Columbia), 77, 86, 101
Vila, Pablo, 155
Vizenor, Gerald, 21–2, 164, 172–3, 175–80, 182–3, 187, 194; *Hotline Healers*, 177, 179, 190–1

Walker, C.P., 78
War of 1812, 204, 226. *See also* Treaty of Ghent
Washington, DC, 18, 134, 216–19
Weaver, Jace, 193
Webster-Ashburton Treaty (1842), 210, 226
Welch, James, 22, 173, 194; *Winter in the Blood*, 175–7, 181, 185, 188–91
West, Elliott, 174–8
Western Hemisphere Travel Initiative, 212–13, 218–19
Westwind Pictures, 42
White Rock (British Columbia), 8, 14–15, 93–4, 96–7, 99–110
WikiLeaks. *See* Department of Homeland Security; Department of State
winkte, 183
Winnipeg (Manitoba), 13, 18, 228, 235;.and early cinema, 71–2, 75–86; floodway, 143. *See also* Lake Winnipeg
Wymond, "Professor," 81–2

xenogamy, 187–91. *See also mestizaje/métissage* and miscegenation
xenophobia, 189

Yaqui nation, 158